Lecture Notes in Control and Information Sciences

Edited by M. Thoma and A. Wyner

For information about Vols. 1–80 please contact your bookseller or Springer-Verlag.

Lecture Notes in Control and Information Sciences

Edited by M. Thoma and A. Wyner

140

Z. Gajić, D. Petkovski,
X. Shen

Singularly Perturbed and Weakly Coupled Linear Control Systems

A Recursive Approach

Springer-Verlag
Berlin Heidelberg GmbH

Authors
Prof. Zoran Gajić
Rutgers University
Dept. of Electrical and Computer Engineering
Piscataway, NJ 08855-0909
USA

Prof. Djordjija Petkovski
University of Novi Sad
Faculty of Technical Sciences
V. Vlahovića 3
21000 Novi Sad
YUGOSLAVIA

Dr. Xuemin Shen
Rutgers University
Dept. of Electrical and Computer Engineering
Piscataway, NJ 08855-0909
USA

ISBN 978-3-540-52333-8   ISBN 978-3-540-46962-9 (eBook)
DOI 10.1007/978-3-540-46962-9

PREFACE

This book is designed to be a fairly comprehensive treatment of the recursive reduced-order methods for singularly perturbed and weakly coupled linear systems. There are numerous examples of singularly perturbed and weakly coupled dynamic systems that provide great challenges to engineers of different disciplines. Obvious examples of singularly perturbed and weakly coupled systems include electrical power systems, aerospace systems, large electric networks, process control systems in chemical and petroleum industries, etc. It is shown that the recursive reduced-order methods are applicable to wider classes of practical problems than the existing singularly perturbed and weakly coupled methods based on the power series expansion. The recursive methods offer several advantages. As it will be shown, the higher order of accuracy can be easily achieved at low cost, the parallel processing of information can be used, results are obtained under much milder assumptions (no analyticity requirements of the problem coefficients), the software and hardware implementation of the control algorithms is highly simplified due to complete parallelism in the design procedures.

This book is intended to the broad audience such as control engineers, applied mathematicians and advanced graduate students who seek a comprehensive view of the current developments in the theory of singularly perturbed and weakly coupled systems. The book emphasizes mathematical developments as well as their application to solving practical problems without assuming strong mathematical background of the readers.

To demonstrate the usefulness of the recursive reduced-order approach to the singularly perturbed and weakly coupled linear systems and to point out its various advantages we have included several real world examples: fluid catalytic cracker, twelve plate absorption column, magnetic type control system, F-8 aircraft, power system composed of two interconnected areas, distillation column, steam power system, and synchronous machine connected to an infinite bus.

We hope that this book will help to reduce some of barriers that exist in recognizing the power and usefulness of the recursive reduced-order methods for singularly perturbed and weakly coupled linear systems, and it will help to broaden their implementation in practice.

Z. Gajić is indebted to his former advisors, Professors H. Khalil and J. Medanić, and to Professor P. Kokotović for bringing him into the challenging research areas of singular perturbations and weak coupling.

Grant support from the following sources is gratefully acknowledged for Dj. Petkovski from the U. S. - Yugoslav Joint Fund for Scientific and Technological Cooperation, in cooperation with the National Science Foundation Grant JF 736, and in cooperation with the Department of Energy under grant JF 727. Dj. Petkovski is particularly thankful to Professor M. Athans and to Dr. A. Levis for fruitful cooperation in the course of these two projects.

The authors are thankful for the contributions of T. Grodt, Professor V. Kecman, N. Harkara, W-C Su, and Đ. Tasevski.

Novi Sad, July 1989.   Authors

TABLE OF CONTENTS

CHAPTER 8.
LINEAR DISCRETE SINGULARLY PERTURBED CONTROL SYSTEMS

INTRODUCTION

Theory of singular perturbations or theory of multiple time scale dynamic systems has been very highly recognized and very rapidly developed control area in the last twenty years (Kokotović and Khalil, 1986, Kokotović, Khalil and O'Reilly, 1986). It has been studied so far from the power series expansion point of view. Being nonrecursive in nature the power series expansion method becomes very cumbersome and computationally very expensive when the high order of accuracy is required. In such cases, the advantage of using the power series expansion method (the important theoretical tool) is questionable from the numerical point of view, and sometimes that method is almost not applicable for practical computations (Grodt and Gajić, 1988, Gajić, Petkovski and Harkara, 1989). In the era of increased application of the modern control theory results to the real world systems that might be a serious problem. In addition, if a small perturbation parameter ε is not very small ("small enough"), then the $O(\varepsilon)$ theory, used so far in the study of singularly perturbed problems, might not produce satisfactory results for the given class of problems. In order to broaden the class of applicable problems the development of the $O(\varepsilon^k)$ theory is a necessary requirement. Even more, it is pointed out in (Hemker, 1983) that the $O(\varepsilon^k)$ theory is the trend in the modern numerical analysis of singularly perturbed problems: "numerical analysis of singular perturbation problems mainly concentrates on the following question: how to find a numerical approximation to the solution for small as well as intermediate values of ε, where no short asymptotic expansion is available. Or, more general, how to construct a single numerical method that can be applied both in the case of extremely small ε and for larger values of ε, when one wouldn't consider the problem as singularly perturbed any longer.".

Furthermore, in the case of singularly perturbed structures induced by a high gain feedback (Kokotović and Khalil, 1986; Kokotović, Khalil and O'Reilly, 1986), the standard statement of the singular perturbation theory "it exists ε small enough" means it exists control input big enough, and thus, that assumption obviously limits practical implementation of the $O(\varepsilon)$ singular perturbation theory quite a lot. In a recent paper (Gajić, Petkovski and Harkara, 1989), a real world example demonstrates a failure of $O(\varepsilon)$ theory for the problem of the optimal static output feedback of linear singularly perturbed systems. The same example is solved successfully in (Gajić, Petkovski and Harkara, 1989) by using $O(\varepsilon^k)$ theory for $k \geq 2$.

The linear weakly coupled systems have been studied in different set-ups by many researchers (Kokotović, Perkins, Cruz and D'Ans, 1969; Delacour, Darwish and Fantin, 1978; Petkovski and Rakić, 1979; Mahmoud, 1978; Sezar and Šiljak, 1986; Ishimatsu, Mohri and Takata, 1975; Washburn and Mendel, 1980; Khalil and Kokotović, 1978). Solutions of the main equations of the linear optimal control theory of weakly coupled systems - Riccati type and/or Lyapunov type — are obtained in terms of the power series expansion of a small coupling parameter ε. Approximate feedback control laws are derived by truncating the expansions of the feedback coefficients of the optimal control law (Kokotović, Perkins, Cruz and D'Ans 1969; Delacour, Darwish and Fantin, 1978; Petkovski and Rakić, 1978; Gajić and Rayavarupu, 1989). Such approximations have been shown to be near-optimal with performance that can be made as close to the optimal performance as desired by including enough terms in the truncated expansions. The recursive approach to weakly coupled systems, based on the fixed point iterations, is developed in (Gajić and Rayavarupu, 1989; Petrović and Gajić, 1988; Harkara, Petkovski and Gajić, 1988; Gajić and Shen, 1989a; Shen and Gajić, 1989a,b,c; Shen, 1989). It has been shown that the recursive methods are particularly useful when the coupling parameter ε is not extremely small and/or when any desired order of accuracy is required, namely, $O(\varepsilon^k)$, where $k = 2, 3, 4, ...$ Even more, in some cases, it is required to achieve very good approximation, such as for a plant-filter augmented system (Shen and Gajić, 1989a), where the accuracy of $O(\varepsilon^k)$, $k \geq 6$ is necessary to stabilize given closed loop system.

This book consists of eight chapters. Chapter 1 comprises an introduction. In Chapter 2 we present a general study of the main algebraic equations of the linear steady state control theory for singularly perturbed and weakly coupled systems, namely, Lyapunov and Riccati equations, and derive corresponding recursive algorithms for their solutions in the most general case when the problem matrices are functions of a small perturbation parameter. The numerical decomposition has been achieved so that only low-order systems are involved in algebraic computations. The introduced recursive methods are of the fixed point type and they are applicable to the wider class of problems than the methods based on the power series expansion. They demand only the boundness of the problem matrices with respect to a small perturbation parameter - contrary to the analyticity requirement of the power series expansion methods. It is shown that the singular perturbation recursive methods converge with the rate of convergence of $O(\varepsilon)$, whereas the recursive methods for weakly coupled linear systems converge faster, that is, with the rate of convergence of $O(\varepsilon^2)$. In the last section of this chapter we introduce a new approach for studying weakly coupled systems via the use of a nonsingular transformation. It completely decouples given weakly coupled system under nonrestrictive assumption. The transformation matrices are obtained from two algebraic matrix equations. Algorithms that efficiently generate solution of these equations are derived.

In Chapter 3 the output feedback control of singularly perturbed and weakly coupled linear systems is studied. Well-defined recursive numerical technique for the solution of nonlinear algebraic matrix equations associated with the output feedback control problem of singularly perturbed systems is developed. The numerical slow-fast decomposition is achieved so that only low-order systems are involved in algebraic computations. It is shown that each iteration step of the proposed algorithm improves the accuracy by an order of magnitude, that is the accuracy of $O(\varepsilon^k)$, can be obtained by performing only k iterations. This represents the significant improvement since all results on the output feedback control problems for the singularly perturbed systems have been obtained so far with the accuracy of $O(\varepsilon)$ only. The real world example, an industrial important reactor - fluid catalytic cracker - demonstrates the efficiency of the proposed algorithm and the failure of $O(\varepsilon)$ theory.

Following similar lines a recursive algorithm is also developed for solving nonlinear algebraic equations comprising the solutions of the optimal static output feedback control problem of linear weakly coupled systems. The effectiveness of the proposed reduced-order algorithm and its advantages over the global full-order algorithm is demonstrated on the twelve plate chemical absorption column. Obtained results strongly support the necessity for the existence of reduced-order numerical techniques for solving corresponding nonlinear algebraic equations. In addition to the reduction in required computations, it can be easier to find a good initial guess and to handle the problem of nonuniqueness of the solution of these nonlinear equations - they represent the necessary conditions only.

In Chapter 4 we present the approach to the decomposition and approximation of the linear-quadratic Gaussian estimation and control problems for weakly coupled systems. The global Kalman filter is decomposed into separate reduced-order local filters via the use of a decopling transformation, introduced in Chapter 2. A near-optimal control law is derived by approximating the coefficients of the optimal control law. The order of approximation of the optimal performance is $O(\varepsilon^k)$, where k is the order of approximation of the coefficients. The electrical power system example demonstrates the failure of $O(\varepsilon^2)$ and $O(\varepsilon^4)$ theory and the necessity for the existence of $O(\varepsilon^6)$ theory. The proposed method produces the reduction in both off-line and on-line computational requirements and it converges under mild assumption. Similarly, in this chapter we study the linear-quadratic Gaussian control problem of singularly perturbed systems. In this context the reduced-order recursive algorithm is used to design a controller for an F-8 aircraft.

Chapter 5 deals with a finite time optimal control problem. In that direction the recursive reduced-order numerical solution of the singularly perturbed and weakly coupled matrix differential Riccati equations are obtained. The order-reductions are achieved in both cases via the use of decoupling transformations applied to the corresponding Hamiltonian matrix. It is shown that corresponding algorithms converge under stabilizability-observability conditions imposed on subsystems with the rate of convergence of $O(\varepsilon)$ for the singularly perturbed and $O(\varepsilon^2)$ for weakly coupled systems. As a case study we present results for the singularly perturbed synchronous machine connected to an infinite bus. The wekly coupled recursive solution of the differential Riccati equation is demonstrated on an example of the distillation column.

The application of the recursive reduced-order approach to differential games is given in Chapter 6. The analysis is restricted to the weakly coupled linear-quadratic Nash games and to the solution of corresponding coupled algebraic Riccati equations. These results can be extended to the other types of differential games either in the context of weakly coupled or singularly perturbed systems.

In Chapters 7 and 8 the discrete time linear systems are studied. The linear weakly coupled discrete systems have not been studied in the literature yet. This is due to the fact that the partitioned expressions of the Riccati equation has a very complicated form in the discrte time domain. We have overcome that problem by the use of the bilinear transformation, which is applicable under quite mild assumption, so that the solution of the discrete algebraic Riccati equation of weakly coupled systems is obtained by using results from Section 2.3.2 derived for the corresonding continuous time Riccati equation. In the remaining part of Chater 7 the obtained results are applied together with the decoupling transformation from Section 2.4, in order to find the near-optimum steady state linear stochastic regulators. As a case study we have considered a fifth order distillation column.

The idea of using the bilinear transformation for the weakly coupled linear systems is extended in Chapter 8 to the singularly perturbed discrete systems. It has been shown that the bilinear transformation preserves the structure of singularly perturbed discrete systems, by retaining slow variables slow and fast variables fast. Then, in the new coordinates the time scale separation can be exploited in order to construct the reduced-order controllers. The linear-quadratic control problem of discrte singularly perturbed systems is solved on a real world example, the F-8 aircraft. The corresponding stochastic problem is studied for a steam power system. The proposed method allows parallel processing of information and it reduces considerably the size of required off-line and on-line computations, since it introduces the full parallelism in the design procedure.

We hope that these results, based on the recursive reduced-reduced order approach, can be extended to the nonlinear singularly perturbed and nonlinear weakly coupled control problems. Research in that direction is under way.

The book is mostly based on the authors recent research papers and we have been following them very closely in many parts of this book: (Gajić and Rayavarupu, 1989, Gajić and Shen, 1989) - Chapter 2, (Gajić, Petkovski and Harkara, 1989, Harkara, Petkovski and Gajić, 1989) - Chapter 3, (Gajić, 1986, Shen and Gajić, 1989a) - Chapter 4, (Grodt and Gajić, 1988, Su and Gajić, 1989) - Chapter 5, (Petrović and Gajić, 1988) - Chapter 6, (Shen and Gajić, 1989b,c) - Chapter 7, and (Gajić and Shen, 1989b,c) - Chapter 8.

ALGEBRAIC LYAPUNOV AND RICCATI EQUATIONS

2.1 Introduction

Development of the recursive techniques for singularly perturbed and weakly coupled linear-quadratic steady state control problems has started recently (Gajić, 1986, Petrović and Gajić, 1988, Gajić, Petkovski and Harkara, 1989, Harkara, Petkovski and Gajić, 1989, Gajić, Rayavarupu, 1989, Gajić and Shen, 1989a,b,c, Shen and Gajić, 1989a,b,c, Shen, 1989). The recursive reduced-order numerical method for a finite time singularly perturbed linear control problem (differential singularly perturbed Riccati equation) is developed in (Grodt and Gajić, 1988). The corresponding weakly coupled differential Riccati equation is studied in (Su and Gajić, 1989). In this chapter we will present a general study of the main algebraic equations of the singularly perturbed linear steady state control theory, namely, Lyapunov and Riccati equations, and derive corresponding recursive, reduced-order parallel algorithms for their solutions in the most general case when the system matrices are functions of a small perturbation parameter. The numerical decomposition has been achieved so that only low-order systems are involved in algebraic computations. The introduced recursive methods are of the fixed point type and they are applicable to the wider class of problems than the methods based on the power series expansion. They demand only the boundness of the problem matrices with respect to a small perturbation parameter - contrary to the analyticity requirement of the power series expansion methods. Similar type of recursive methods will be developed for the algebraic Lyapunov and Riccati equations of the weakly coupled linear systems. In addition, in the last section in this chapter we will develop a nonsingular transformation that completely decouples a linear weakly coupled system into two independent subsystems.

The proposed methods allow parallel processing of information and they reduce considerably the size of required computations, since they introduce the full parallelism in the design procedures.

It is shown that developed reduced-order parallel algorithms converge to the required solutions with the rate of convergence of $O(\varepsilon^2)$ for weakly coupled systems, and only with $O(\varepsilon)$ for the singularly perturbed systems.

2.2 The Recursive Methods for Singularly Perturbed Linear Systems

Consider a linear system

$$\dot{x} = A(\varepsilon)x + B(\varepsilon)u, \qquad x(0) = x_0 \tag{2.1}$$

with a performance index

$$J(\varepsilon) = \frac{1}{2} \int_0^\infty [x^T Q(\varepsilon)x + u^T R(\varepsilon)u]dt, \qquad Q(\varepsilon) \geq 0, \quad R(\varepsilon) > 0 \tag{2.2}$$

which has to be minimized, and where ε is a small parameter (it is a positive one for singularly perturbed systems and of arbitrary sign for weakly coupled systems). $x \in R^n$ is a state vector, and $u \in R^m$ is an input vector, and all matrices are of appropriate dimensions. The optimal control $u(t)$ that minimizes (2.2) along trajectories of (2.1) is given by the well known expression

$$u(t) = -R^{-1}(\varepsilon)B^T(\varepsilon)P(\varepsilon)x(t) \tag{2.3}$$

where $P(\varepsilon)$ is the positive semidefinite stabilizing solution of the algebraic Riccati equation

$$P(\varepsilon)A(\varepsilon) + A^T(\varepsilon)P(\varepsilon) + Q(\varepsilon) - P(\varepsilon)S(\varepsilon)P(\varepsilon) = 0, \quad S(\varepsilon) = B(\varepsilon)R^{-1}(\varepsilon)B^T(\varepsilon) \tag{2.4}$$

For $S(\varepsilon) = 0$, the equation (2.4) becomes the algebraic Lyapunov equation. In this chapter we will also study a dual form of the algebraic Lyapunov equation that represents a variance equation of a linear system driven by white noise

$$\dot{x} = A(\varepsilon)x + G(\varepsilon)w \qquad (2.5)$$

where w is a zero-mean Gaussian white noise with a unity intensity matrix. The algebraic Lyapunov equation corresponding to (2.5) is given by

$$K(\varepsilon)A^T(\varepsilon) + A(\varepsilon)K(\varepsilon) + G(\varepsilon)G^T(\varepsilon) = 0 \qquad (2.6)$$

According to the theory of singular perturbations (Kokotović and Khalil, 1986, Kokotović, Khalil and O'Reilly 1986), the following partitions of the problem matrices are introduced

$$A(\varepsilon) = \begin{bmatrix} A_1(\varepsilon) & A_2(\varepsilon) \\ A_3(\varepsilon)/\varepsilon & A_4(\varepsilon)/\varepsilon \end{bmatrix}, \quad B(\varepsilon) = \begin{bmatrix} B_1(\varepsilon) \\ B_2(\varepsilon)/\varepsilon \end{bmatrix}, \quad G(\varepsilon) = \begin{bmatrix} G_1(\varepsilon) \\ G_2(\varepsilon)/\varepsilon \end{bmatrix}$$

$$P(\varepsilon) = \begin{bmatrix} P_1(\varepsilon) & \varepsilon P_2(\varepsilon) \\ \varepsilon P_2^T(\varepsilon) & \varepsilon P_3(\varepsilon) \end{bmatrix}, \quad K(\varepsilon) = \begin{bmatrix} K_1(\varepsilon) & K_2(\varepsilon) \\ K_2^T(\varepsilon) & K_3(\varepsilon)/\varepsilon \end{bmatrix} \qquad (2.7)$$

$$Q(\varepsilon) = \begin{bmatrix} Q_1(\varepsilon)\,Q_1^T(\varepsilon) & Q_1(\varepsilon)\,Q_2^T(\varepsilon) \\ Q_2(\varepsilon)\,Q_1^T(\varepsilon) & Q_2(\varepsilon)\,Q_2^T(\varepsilon) \end{bmatrix}$$

Newly defined matrices are of dimensions A_1, P_1, K_1, $Q_1 \in R^{n_1 \times n_1}$; A_4, P_3, K_3, $Q_3 \in R^{n_2 \times n_2}$, B_i, $G_i \in R^{n_i \times m}$, $i = 1, 2$, with $n_1 + n_2 = n$. It is assumed that all matrices are continuous functions of ε.

2.2.1 The Recursive Reduced-Order Method for the Algebraic Lyapunov Equation

The partitioned form of the Lyapunov equation defined in (2.6) is

$$A_1(\varepsilon)K_1(\varepsilon) + K_1(\varepsilon)A_1^T(\varepsilon) + A_2(\varepsilon)K_2^T(\varepsilon) + K_2(\varepsilon)A_2^T(\varepsilon) + G_1(\varepsilon)G_1^T(\varepsilon) = 0$$

$$K_2(\varepsilon)A_4^T(\varepsilon) + \varepsilon A_1(\varepsilon)K_2(\varepsilon) + K_1(\varepsilon)A_3^T(\varepsilon) + A_2(\varepsilon)K_3(\varepsilon) + G_1(\varepsilon)G_2^T(\varepsilon) = 0 \qquad (2.8)$$

$$K_3(\varepsilon)A_4^T(\varepsilon) + A_4(\varepsilon)K_3(\varepsilon) + \varepsilon A_3(\varepsilon)K_2^T(\varepsilon) + \varepsilon K_2^T(\varepsilon)A_3^T(\varepsilon) + G_2(\varepsilon)G_2^T(\varepsilon) = 0$$

Let us define the following $O(\varepsilon)$ perturbation of (2.8)

$$A_1(\varepsilon)K_1(\varepsilon) + K_1(\varepsilon)A_1^T(\varepsilon) + K_2(\varepsilon)A_2^T(\varepsilon) + A_2(\varepsilon)K_2^T(\varepsilon) + G_1(\varepsilon)G_1^T(\varepsilon) = 0$$

$$A_2(\varepsilon)K_3(\varepsilon) + K_2(\varepsilon)A_4^T(\varepsilon) + K_1(\varepsilon)A_3^T(\varepsilon) + G_1(\varepsilon)G_2^T(\varepsilon) = 0 \qquad (2.9)$$

$$K_3(\varepsilon)A_4^T(\varepsilon) + A_4(\varepsilon)K_3(\varepsilon) + G_2(\varepsilon)G_2^T(\varepsilon) = 0$$

Note that we did not set $\varepsilon = 0$ in A_i's and G_i's. In the rest of the chapter we will assume that all matrices are functions of ε. However, the explicit dependence on ε of the problem matrices will be omitted in order to simplify notation. Solution of (2.9) is in fact given in terms of two lower-order algebraic Lyapunov equations

$$A_0 K_1 + K_1 A_0^T + G_0 G_0^T = 0$$
$$\qquad (2.10)$$
$$A_4 K_3 + K_3 A_4^T + G_2 G_2^T = 0$$

and

$$K_2 = - \left[A_2 K_3 + K_1 A_3^T + G_1 G_2^T \right] A_4^{-T} \qquad (2.11)$$

where

$$A_0 = A_1 - A_2 A_4^{-1} A_3, \quad G_0 = G_1 - A_2 A_4^{-1} G_2 \qquad (2.12)$$

Unique solutions of (2.10)-(2.11) exist under the following assumption.

Assumption 2.1. Matrices $A_0(\varepsilon)$ and $A_4(\varepsilon)$ are stable.

This is a standard assumption in the theory of singular perturbations (Kokotović and Khalil, 1986, Kokotović, Khalil and O'Reilly, 1986). Defining approximation errors as

$$K_1 = \bar{K}_1 + \varepsilon E_1$$

$$K_2 = \bar{K}_2 + \varepsilon E_2 \qquad (2.13)$$

$$K_3 = \bar{K}_3 + \varepsilon E_3$$

and subtracting (2.10)-(2.11) from (2.9), we get the error equations (after some algebra) in the form

$$A_0 E_1 + E_1 A_0^T = A_0 \left[K_2 + \varepsilon E_2 \right] A_4^{-T} A_2^T + A_2 A_4^{-1} \left[K_2 + \varepsilon E_2 \right]^T A_0^T$$

$$A_4 E_3 + E_3 A_4^T = -A_3 \left[K_2 + \varepsilon E_2 \right] - \left[K_2 + \varepsilon E_2 \right]^T A_3^T \qquad (2.14)$$

$$A_2 E_3 + E_1 A_3^T + E_2 A_4^T + A_1 \left[K_2 + \varepsilon E_2 \right] = 0$$

These equations have very nice forms since the unknown quantity E_2 in equations for E_1 and E_3 is multiplied by a small parameter ε. This fact suggests the following reduced-order parallel algorithm for solving (2.14)

$$A_0 E_1^{(i+1)} + E_1^{(i+1)} A_0^T = A_0 \left[K_2 + \varepsilon E_2^{(i)} \right] A_4^{-T} A_2^T + A_2 A_4^{-1} \left[K_2 + \varepsilon E_2^{(i)} \right]^T A_0^T$$

$$A_4 E_3^{(i+1)} + E_3^{(i+1)} A_4^T = -A_3 \left[K_2 + \varepsilon E_2^{(i)} \right] - \left[K_2 + \varepsilon E_2^{(i)} \right]^T A_3^T \qquad (2.15)$$

$$E_2^{(i+1)} = - \left\{ A_2 E_3^{(i+1)} + E_1^{(i+1)} A_3^T + A_1 \left[K_2 + \varepsilon E_2^{(i)} \right] \right\} A_4^{-T} \quad i = 0, 1, 2, \dots$$

with the starting point $E_2^{(0)} = 0$.

Using the stability Assumption 2.1, it is easy to show that

$$\| E_j^{(i)} - E_j \| = O(\varepsilon^i), \qquad j = 1, 2, 3, \quad i = 1, 2, \dots \qquad (2.16)$$

Note that (2.16) is valid even in the case when the last equation of (2.15) is in the form

$$E_2^{(i+1)} = - \{A_2 E_3^{(i)} + E_1^{(i)} A_3^T + A_1 [K_2 + \varepsilon E_2^{(i)}]\} A_4^{-T}, \quad i = 0, 1, 2, \ldots \quad (2.17)$$

with $E_1^{(0)} = 0$ and $E_3^{(0)} = 0$.

Thus, the algorithm (2.15) is convergent. Using $E_j^{(\infty)}$, $j = 1, 2, 3$ in (2.15) and comparing it to (2.14), imply that the algorithm (2.15) converges to the unique solution of (2.14). In summary we have the following theorem.

Theorem 2.1 Under stability assumptions imposed on $A_0(\varepsilon)$ and $A_4(\varepsilon)$, the algorithm (2.15) converges to the exact solution E with the rate of convergence of $O(\varepsilon)$, and thus, the required solution K can be obtained with the accuracy of $O(\varepsilon^i)$ from

$$K_j^{(i)} = K_j + \varepsilon E_j^{(i)} = K_j + O(\varepsilon^i), \quad j = 1, 2, 3, \quad i = 1, 2, \ldots \quad (2.18)$$

It is important to notice that in the proposed method we do not need to expand $A_i(\varepsilon)$, $i = 1, \ldots, 4$ into power series, and we do not require stability of $A_0(0)$ and $A_4(0)$ which make the important features of the presented method. It is known that the power series expansion method leads to two reduced-order Lyapunov equations similar to those in (2.15) — they are of the same order, but the number of terms on the right hand side of these equations for the power series expansion method is growing very quickly with an increase in the required accuracy. It can be seen from (2.15) that for the fixed point method the number of terms on the right hand side is constant. The number of matrix multiplications required to form right hand sides of Lyapunov equations, corresponding to the fast variables, for the accuracy of $O(\varepsilon^i)$, is given in Table 2.1

i	1	2	3	4	5	6
fixed point	1	1	1	1	1	1
power series	3	6	9	12	15	18

Table 2.1 Required number of matrix multiplications

This table shows very strong support for the proposed fixed point method.

Assumption 2.1 is much more natural and less binding than the stability assumption imposed on $A_0(0)$ and $A_4(0)$, (Kokotović and Khalil, 1986, Kokotović, Khalil and O'Reilly, 1986). Namely, the singularly perturbed structure of the system is the consequence of a strict inequality $\varepsilon > 0$ (small positive parameter). Requirement imposed on $A_0(0)$ and $A_4(0)$ is based on the continuation argument, but it can not be indefinitely exploited. It is possible that, for example, $A_4(\varepsilon)$ - feedback matrix of a real singularly perturbed system - is stable, but $A_4(0)$ is unstable. Let us take a very simple scalar example, where $A_4(\varepsilon) = 0.2 - \varepsilon$. This matrix is stable for all $\varepsilon > 0.2$. Thus for any $\varepsilon \in (0.2, \varepsilon_{max})$, where ε_{max} is obtained from the requirement that the radius of convergence of the fixed point algorithm is less than 1, the proposed algorithm would be convergent, in spite of the fact that $A_4(0)$ is the unstable matrix.

In addition, the power series method demands analyticity of all problem matrices with respect to ε at $\varepsilon = 0$, whereas for the application of the fixed point method, we need only continuity of the same matrices (Zangwill and Garcia, 1981).

Finally, the main advantage of the presented fixed point algorithm is in its parallel and recursive structure.

2.2.2 The Recursive Reduced-Order Method for the Algebraic Riccati Equation

This approach was first developed in (Gajić, 1986) for the non-parametrized case. In this section we will study the fixed point method to the solution of the algebraic Riccati equation for a more general case when the problem matrices are continuous functions of ε.

Consider the algebraic Riccati equation of singularly perturbed systems defined in (2.4) and (2.7). Partitioning (2.4) subject to (2.7) we get the following equations

$$P_1A_1 + A_1^TP_1 + P_2A_3 + A_3^TP_2^T - P_1S_1P_1 - P_1SP_2^T - P_2S^TP_1$$

$$- P_2S_2P_2^T + Q_1^TQ_1 = 0 \tag{2.19}$$

$$P_1A_2 + P_2A_4 + \varepsilon A_1^TP_2 + A_3^TP_3 - \varepsilon P_1S_1P_2 - P_1SP_3 - \varepsilon P_2S^TP_2$$

$$- P_2S_2P_3 + Q_1^TQ_2 = 0 \tag{2.20}$$

$$P_3A_4 + A_4^TP_3 + \varepsilon P_2^TA_2 + \varepsilon A_2^T P_2 - P_3S_2P_3 - \varepsilon^2 P_2^TS_1P_2 - \varepsilon P_2^TSP_3$$

$$- \varepsilon P_3S^TP_2 + Q_2^TQ_2 = 0 \tag{2.21}$$

where

$$S_i = B_iR^{-1}B_i^T , \; i = 1, 2, \qquad S = B_1R^{-1}B_2^T \tag{2.22}$$

Let us define the following $O(\varepsilon)$ perturbation of (2.19)-(2.21)

$$P_1A_1 + A_1^TP_1 + P_2A_3 + A_3^TP_2^T - P_1S_1P_1 -$$

$$P_1SP_2^T - P_2S^TP_1 - P_2S_2P_2^T + Q_1^TQ_1 = 0 \tag{2.23}$$

$$P_1A_2 + P_2A_4 + A_3^TP_3 - P_1SP_3 - P_2S_2P_3 + Q_1^TQ_2 = 0 \tag{2.24}$$

$$P_3A_4 + A_4^TP_3 - P_3S_2P_3 + Q_2^TQ_2 = 0 \tag{2.25}$$

It is important to point out that ε in the coefficient matrices is not set to zero.

The Riccati equation (2.25) will produce the unique positive semidefinite stabilizing solution under the following assumption.

Assumption 2.2. The triple $(A_4(\varepsilon), B_2(\varepsilon), Q_2(\varepsilon))$ is stabilizable and detectable.

From (2.24) we obtain

$$P_2 = - (P_1A_2 + A_3^TP_3 - P_1SP_3 + Q_1^TQ_2) (A_4 - S_2P_3)^{-1} \tag{2.26}$$

which after a substitution in (2.23) and elimination of P_3 produces the reduced-order slow algebraic Riccati equation, (Kokotović and Khalil, 1986, Kokotović, Khalil and O'Reilly, 1986), in the form

$$P_1A + A^TP_1 - P_1SP_1 + Q = 0 \tag{2.27}$$

where

$$A = A_0 - B_0R_0^{-1}r^TQ_0, \qquad S = B_0R_0^{-1}B_0^T, \qquad B_0 = B_1 - A_2A_4^{-1}B_2$$

$$Q_0 = Q_1 - Q_2A_4^{-1}A_3, \qquad R_0 = R + r^Tr, \qquad r = -Q_2A_4^{-1}B_2 \tag{2.28}$$

$$Q = Q_0^T(I - rR_0^{-1}r^T)Q_0$$

The unique positive semidefinite stabilizing solution of (2.27) exists under the following assumption.

Assumption 2.3 The triple $\left(A(\varepsilon), B_0(\varepsilon), \sqrt{Q(\varepsilon)}\right)$ is stabilizable and detectable.

Therefore the zero-order solution has the form

$$P(\varepsilon) = \begin{bmatrix} P_1(\varepsilon) & \varepsilon P_2(\varepsilon) \\ \varepsilon P_2^T(\varepsilon) & \varepsilon P_3(\varepsilon) \end{bmatrix} \tag{2.29}$$

The zero-order solution is $O(\varepsilon)$ close to the exact one. We define errors as

$$P_j(\varepsilon) = P_j(\varepsilon) + \varepsilon E_j(\varepsilon), \qquad j = 1, 2, 3 \tag{2.30}$$

The $O(\varepsilon^k)$ approximation of E_1's will produce the $O(\varepsilon^{k+1})$ approximation of the required matrix P, which is why we are interested in finding equations for the error term and a convenient algorithm for its solution. Subtracting (2.23)-(2.25) from (2.19)-(2.21) and using (2.30) we arrive at the following expression for the error equation

$$E_1 D_1 + D_1^T E_1 = D^T H_1 + H_1 D + D^T H_3 D + \epsilon H_2$$

$$E_2 D_3 + E_1 D_{21} + D_{22}^T E_3 = -H_1 \tag{2.31}$$

$$E_3 D_3 + D_3^T E_3 = H_3$$

where

$$D_3 = A_4 - S_2 P_3, \quad D_{22} = A_3 - S^T P_1 - S_2 P_2^T$$

$$D_{21} = A_2 - SP_3, \quad D = D_3^{-1} D_{22} \tag{2.32}$$

$$D_{11} = A_1 - S_1 P_1 - SP_2^T, \quad D_1 = D_{11} - D_{21} D_3^{-1} D_{22}$$

and

$$H_1 = A_1^T P_2 - P_1 S_1 P_2 - P_2 S^T P_2 - \epsilon(E_1 SE_3 + E_2 S_2 E_3)$$

$$H_2 = E_1 S_1 E_1 + E_1 SE_2^T + E_2 S^T E_1 + E_2 S_2 E_2^T \tag{2.33}$$

$$H_3 = -P_2^T A_2 - A_2^T P_2 + \epsilon P_2^T S_1 P_2 + \epsilon E_3 S_2 E_3 + P_2^T SP_3 + P_3 S^T P_2$$

Equations (2.31) have all cross-coupling terms and all nonlinear terms multiplied by a small parameter ε, which suggests that a fixed point algorithm can be efficient for their solution. We propose the following algorithm, similar to one obtained in (Gajić, 1986) for the nonparametrized case

$$E_3^{(i+1)} D_3 + D_3^T E_3^{(i+1)} = H_3^{(i)}$$

$$E_2^{(i+1)} D_3 + E_1^{(i+1)} D_{21} + D_{22}^T E_3^{(i+1)} = -H_1^{(i)} \tag{2.34}$$

$$E_1^{(i+1)} D_1 + D_1^T E_1^{(i+1)} = D^T H_1^{(i)T} + H_1^{(i)} D + D^T H_3^{(i)} D + \epsilon H_2^{(i)}$$

$$E_1^{(0)} = 0, \quad E_2^{(0)} = 0, \quad E_3^{(0)} = 0, \qquad i = 0, 1, 2, 3, \ldots$$

The following theorem indicates the features of the algorithm (2.34).

Theorem 2.2. Under stabilizability-detectability conditions, imposed in Assumptions 2.2 and 2.3, the algorithm (2.34) converges to the exact solution of E with the rate of convergence of $O(\varepsilon)$, that is

$$\| E - E^{(i+1)} \| = O(\varepsilon) \| E - E^{(i)} \| , \qquad i = 0, 1, 2, 3, ... \qquad (2.35)$$

or equivalently

$$\| E - E^{(i)} \| = O(\varepsilon^i) \qquad (2.36)$$

Proof: As a starting point we need to show the existence of bounded solutions of E_1, E_2, and E_3, in the neighbourhood of ε^*, where $\varepsilon^* \in [\varepsilon_{min}, \varepsilon_{max}]$. To prove that, by the implicit function theorem, it is enough to show, that the corresponding Jacobian is nonsingular at ε^*. The Jacobian is given by

$$J(\varepsilon) = \begin{bmatrix} J_{11}(\varepsilon) & 0 & 0 \\ J_{21}(\varepsilon) & J_{22}(\varepsilon) & J_{23}(\varepsilon) \\ 0 & 0 & J_{33}(\varepsilon) \end{bmatrix} + \begin{bmatrix} 0 & 0 & 0 \\ O(\varepsilon) & O(\varepsilon) & 0 \\ 0 & O(\varepsilon) & O(\varepsilon) \end{bmatrix} \qquad (2.37)$$

Using the Kronecker product representation we have

$$J_{11} = D_{11}{}^T \oplus I_{n_1} + I_{n_1} \oplus D_{11}{}^T$$

$$J_{33} = D_3{}^T \oplus I_{n_2} + I_{n_2} \oplus D_3{}^T \qquad (2.38)$$

$$J_{22} = D_3{}^T \oplus I_{n_2}$$

For the Jacobian to be nonsingular J_{ii}, $i = 1, 2, 3$ have to be nonsingular. Matrix D_3 is a closed loop matrix of the fast subsystem, and thus stable by the well known properties of the solution of the algebraic Riccati equation. Matrix D_1 can be easily shown to be the closed loop matrix of the reduced slow subsystem and it is stable also (Kokotović and Khalil, 1986, Kokotović, Khalil and O'Reilly, 1986). By known property of the Kronecker product, (Lancaster and Tismenetsky, 1985), matrices J_{ii} are then nonsingular. Thus, for ε^* small enough the Jacobian $J(\varepsilon^*)$ is nonsingular. The second part of the proof is to produce an estimate of the rate of convergence and to verify (2.35) and (2.36). That can be done similarly to (Gajić, 1986), and thus, it is omitted.

Therefore we are able to find the exact solution of the full order algebraic Riccati equation of singularly perturbed systems, by recursively solving two reduced-order Lyapunov equations and one linear equation in a parallel manner.

Numerical examples, in different set-ups, that demonstrate the efficiency of the presented algorithms for solving algebraic Lyapunov and Riccati equations of singularly perturbed systems will be given in next chapters.

2.3 The Recursive Methods for Weakly Coupled Linear Systems

The weakly coupled linear-quadratic control problem is defined by (2.1)-(2.4) subject to the following partition of the problem matrices (Kokotović, Perkins, Cruz and D'Ans, 1969)

$$
A(\varepsilon) = \begin{bmatrix} A_1(\varepsilon) & \varepsilon A_2(\varepsilon) \\ \varepsilon A_3(\varepsilon) & A_4(\varepsilon) \end{bmatrix}, \qquad B(\varepsilon) = \begin{bmatrix} B_1(\varepsilon) & \varepsilon B_2(\varepsilon) \\ \varepsilon B_3(\varepsilon) & B_4(\varepsilon) \end{bmatrix}
$$

$$
Q(\varepsilon) = \begin{bmatrix} Q_1(\varepsilon) & \varepsilon Q_2(\varepsilon) \\ \varepsilon Q_2^T(\varepsilon) & Q_3(\varepsilon) \end{bmatrix}, \qquad R(\varepsilon) = \begin{bmatrix} R_1(\varepsilon) & 0 \\ 0 & R_2(\varepsilon) \end{bmatrix}
$$

(2.39)

where ε is a small parameter. Dimensions of partitioned matrices are compatible to those defined in (2.7). In this section we will study the recursive fixed point type parallel algorithms for solving algebraic Lyapunov and Riccati equations of weakly coupled systems.

2.3.1 The Recursive Reduced-Order Parallel Algorithm for Solving the Algebraic Lyapunov Equation of Weakly Coupled Systems

The algebraic Lyapunov equation of weakly coupled systems ("regulator type") is given by

$$A^T(\varepsilon)P(\varepsilon) + P(\varepsilon)A(\varepsilon) + Q(\varepsilon) = 0 \tag{2.40}$$

Due to block dominant structure of matrices A and Q, the required solution P is properly scaled as follows

$$P(\varepsilon) = \begin{bmatrix} P_1(\varepsilon) & \varepsilon P_2(\varepsilon) \\ \varepsilon P_2^T(\varepsilon) & P_3(\varepsilon) \end{bmatrix} \tag{2.41}$$

Partitioned form of (2.40) produces

$$P_1 A_1 + A_1^T P_1 + Q_1 + \varepsilon^2 (P_2 A_3 + A_3^T P_2^T) = 0$$

$$P_1 A_2 + P_2 A_4 + A_1^T P_2 + A_3^T P_3 + Q_2 = 0 \tag{2.42}$$

$$P_3 A_4 + A_4^T P_3 + Q_3 + \varepsilon^2 (P_2^T A_2 + A_2^T P_2) = 0$$

We define the $O(\varepsilon^2)$ approximation of (2.42) as

$$P_1 A_1 + A_1^T P_1 + Q_1 = 0$$

$$P_2 A_4 + A_1^T P_2 = - P_1 A_2 - A_3^T P_3 - Q_2 \tag{2.43}$$

$$P_3 A_4 + A_4^T P_3 + Q_3 = 0$$

Note that we did not set $\varepsilon = 0$ in A_i's and Q_i's, so that P_i's are functions of ε.

The unique solution of (2.43) exists under the following assumption.

Assumption 2.4. Matrices $A_1(\varepsilon)$ and $A_4(\varepsilon)$ are stable.

Defining approximation errors as

$$P_j = P_j + \varepsilon^2 E_j , \qquad j = 1, 2, 3 \tag{2.44}$$

and subtracting (2.43) from (2.42) we obtain the following expression for the errors

$$E_1 A_1 + A_1^T E_1 + P_2 A_3 + A_3^T P_2^T + \varepsilon^2\left(E_2 A_3 + A_3^T E_2^T\right) = 0$$

$$E_2 A_4 + A_1^T E_2 + E_1 A_2 + A_3^T E_3 = 0 \tag{2.45}$$

$$E_3 A_4 + A_4^T E_3 + A_2^T P_2 + P_2^T A_2 + \varepsilon^2\left(A_2^T E_2 + E_2^T A_2\right) = 0$$

We propose the following algorithm, having the reduced-order and parallel structure, for solving (2.45)

$$E_1^{(i+1)} A_1 + A_1^T E_1^{(i+1)} + P_2^{(i)} A_3 + A_3^T P_2^{(i)^T} = 0$$

$$E_3^{(i+1)} A_4 + A_4 E_3^{(i+1)} + A_2^T P_2^{(i)} + P_2^{(i)^T} A_2 = 0 \tag{2.46}$$

$$E_2^{(i+1)} A_4 + A_1^T E_2^{(i+1)} + E_1^{(i+1)} A_2 + A_3^T E_1^{(i+1)} = 0, \quad i = 0, 1, 2, \dots$$

with the starting point $E_2^{(0)} = 0$ and with

$$P_j^{(i)} = P_j + \varepsilon E_j^{(i)}, \qquad j = 1, 2, 3, \qquad i = 0, 1, 2, \dots \tag{2.47}$$

Using the same arguments like in Section 2.2, we can establish the following theorem.

Theorem 2.3. Under stability assumptions imposed on matrices $A_1(\varepsilon)$ and $A_4(\varepsilon)$, the algorithm (2.46) converges to the exact solution E with the rate of convergence of $O(\varepsilon^2)$, and thus, the required solution P can be obtained with the accuracy of $O(\varepsilon^{2i})$ from (2.47), that is

$$P_j = P_j^{(i)} + O(\varepsilon^{2i}), \quad j = 1, 2, 3, \quad i = 1, 2, 3, \dots \tag{2.48}$$

2.3.2 The Recursive Reduced-Order Parallel Algorithm for Solving the Algebraic Riccati Equation of Weakly Coupled Systems

The algebraic Riccati equation (2.4), subject to the weakly coupled structure given in (2.39), has the solution partitioned as in (2.41). Using (2.39) and (2.41) in (2.4) will produce the following partitioned equations

$$P_1 A_1 + A_1^T P_1 + Q_1 - P_1 S_1 P_1 + \epsilon^2 (P_2 A_3 + A_3^T P_2^T)$$
$$- \epsilon^2 \left[(P_1 S_{12} + P_2 Z^T) P_1 + (P_1 Z + P_2 (S_2 + \epsilon^2 S_{21})) P_2^T \right] = 0 \tag{2.49}$$

$$P_3 A_4 + A_4^T P_3 + \epsilon^2 (P_2^T A_2 + A_2^T P_2) + Q_3 - P_3 (S_2 + \epsilon^2 S_{21}) P_3$$
$$- \epsilon^2 \left\{ [P_2^T (S_1 + \epsilon^2 S_{12}) + P_3 Z^T] P_2 + P_2^T Z P_3 \right\} = 0 \tag{2.50}$$

$$P_1 A_2 + P_2 A_4 + A_1^T P_2 + A_3^T P_3 + Q_2 - P_1 S_1 P_2 - P_1 Z P_3 - P_2 S_2 P_3$$
$$- \epsilon^2 \left[(P_1 S_{12} + P_2 Z^T) P_2 + P_2 S_{21} P_3 \right] = 0 \tag{2.51}$$

where

$$S_1 = B_1 R_1^{-1} B_1^T, \quad S_2 = B_4 R_2^{-1} B_4^T, \quad S_{12} = B_2 R_2^{-1} B_2^T$$
$$S_{21} = B_3 R_1^{-1} B_3^T, \quad Z = B_1 R_1^{-1} B_3^T + B_2 R_2^{-1} B_4^T \tag{2.52}$$

The $O(\epsilon^2)$ approximation of (2.49)-(2.51) is defined as

$$P_1 A_1 + A_1^T P_1 - P_1 S_1 P_1 + Q_1 = 0$$
$$P_3 A_4 + A_4^T P_3 - P_3 S_2 P_3 + Q_3 = 0 \tag{2.53}$$

and

$$P_2 D_2 + D_1^T P_2 = - (P_1 A_2 + A_3^T P_3 + Q_2 - P_1 Z P_3) \tag{2.54}$$

where

$$D_1(\varepsilon) = \left[A_1(\varepsilon) - S_1(\varepsilon)P_1(\varepsilon)\right], \qquad D_2(\varepsilon) = \left[A_4(\varepsilon) - S_2(\varepsilon)P_3(\varepsilon)\right] \qquad (2.55)$$

The unique positive semidefinite stabilizing solutions of (2.53) exist under the following assumption.

Assumption 2.5. Triples $\left(A_1(\varepsilon), B_1(\varepsilon), \sqrt{Q_1}(\varepsilon)\right)$ and $\left(A_4(\varepsilon), B_4(\varepsilon), \sqrt{Q_3}(\varepsilon)\right)$ are stabilizable-detectable.

Under this assumption matrices $D_1(\varepsilon)$ and $D_2(\varepsilon)$ are stable so that the unique solution of (2.54) exist also. If the errors are defined as

$$P_j = P_j + \varepsilon^2 E_j, \quad j = 1, 2, 3 \qquad (2.56)$$

then the exact solution will be of the form

$$P = \begin{bmatrix} P_1 + \varepsilon^2 E_1 & \varepsilon\left(P_2 + \varepsilon^2 E_2\right) \\ \varepsilon\left(P_2 + \varepsilon^2 E_2\right)^T & P_3 + \varepsilon^2 E_3 \end{bmatrix} \qquad (2.57)$$

Subtracting (2.53) and (2.54) from the corresponding equations (2.49)-(2.51) and using (2.56) produce the following equations for the errors

$$E_1 D_1 + D_1^T E_1 = P_1 S_{12} P_1 + P_2 Z^T P_1 + P_1 Z P_2^T + P_2 S_2 P_2^T \qquad (2.58)$$

$$-P_2 A_3 - A_3^T P_2^T + \varepsilon^2 \left(E_1 S_1 E_1 + P_2 S_{21} P_2^T\right)$$

$$E_3 D_2 + D_2^T E_3 = P_3 S_{21} P_3 + P_2^T S_1 P_2 + P_3 Z^T P_2 + P_2^T Z P_3 \qquad (2.59)$$

$$- P_2^T A_2 - A_2^T P_2 + \varepsilon^2 \left(E_3 S_2 E_3 + P_2^T S_{12} P_2\right)$$

$$D_1^T E_2 + E_2 D_2 = P_1 S_{12} P_2 + P_2 Z^T P_2 + P_2 S_{21} P_3 \qquad (2.60)$$

$$- E_1 D_{12} - D_{21}^T E_3 + \varepsilon^2 \left(E_1 S_1 E_2 + E_1 Z E_2 + E_2 S_2 E_3\right)$$

where

$$D_{12} = A_2 - S_1 P_2 - Z P_3, \quad D_{21} = A_3 - S_2 P_2^T - Z^T P_1 \qquad (2.61)$$

It can be easily shown that the nonlinear equations (2.58)-(2.60) have the form as

$$E_1 D_1 + D_1{}^T E_1 = \text{const} + \varepsilon^2 f_1\left(E_1, E_2, \varepsilon^2\right)$$

$$E_3 D_2 + D_2{}^T E_3 = \text{const} + \varepsilon^2 f_3\left(E_2, E_3, \varepsilon^2\right) \tag{2.62}$$

$$E_2 D_2 + D_1{}^T E_2 = \text{const} + \varepsilon^2 f_2\left(E_1, E_2, E_3, \varepsilon^2\right)$$

We can see that all cross coupling terms and all nonlinear terms in (2.58)-(2.60) are multiplied by ε^2, so that we propose the following reduced-order parallel algorithm for solving (2.58)-(2.60)

$$
\begin{aligned}
E_1{}^{(i+1)} D_1 + D_1{}^T E_1{}^{(i+1)} &= P_1{}^{(i)} S_{12} P_1{}^{(i)} + P_2{}^{(i)} Z^T P_1{}^{(i)} + P_1{}^{(i)} Z P_2{}^{(i)T} \\
&+ P_2{}^{(i)} S_2 P_2{}^{(i)T} - P_2{}^{(i)} A_3 - A_3{}^T P_2{}^{(i)T} + \varepsilon^2 \left(E_1{}^{(i)} S_1 E_1{}^{(i)} + P_2{}^{(i)} S_{21} P_2{}^{(i)T} \right)
\end{aligned}
\tag{2.63}
$$

$$
\begin{aligned}
E_3{}^{(i+1)} D_2 + D_2{}^T E_3{}^{(i+1)} &= P_3{}^{(i)} S_{21} P_3{}^{(i)} + P_2{}^{(i)T} S_1 P_2{}^{(i)} + P_3{}^{(i)} Z^T P_2{}^{(i)} \\
&+ P_2{}^{(i)T} Z P_3{}^{(i)} - P_2{}^{(i)T} A_2 - A_2{}^T P_2{}^{(i)} + \varepsilon^2 \left(E_3{}^{(i)} S_2 E_3{}^{(i)} + P_2{}^{(i)T} S_{12} P_2{}^{(i)} \right)
\end{aligned}
\tag{2.64}
$$

$$
\begin{aligned}
D_1{}^T E_2{}^{(i+1)} + E_2{}^{(i+1)} D_2 &= P_1{}^{(i+1)} S_{12} P_2{}^{(i)} + P_2{}^{(i)} Z^T P_2{}^{(i)} + P_2{}^{(i)} S_{21} P_3{}^{(i)} \\
&- E_1{}^{(i+1)} D_{12} - D_{21} T E_3{}^{(i+1)} + \varepsilon^2 \left(E_1{}^{(i+1)} S_1 E_2{}^{(i)} + E_1{}^{(i+1)} Z E_2{}^{(i)} + E_2{}^{(i)} S_2 E_3{}^{(i+1)} \right)
\end{aligned}
\tag{2.65}
$$

with $E_1{}^{(0)} = 0$, $E_2{}^{(0)} = 0$, $E_3{}^{(0)} = 0$, where

$$P_j{}^{(i)} = P_j + \varepsilon^2 E_j{}^{(i)}, \ j = 1, 2, 3, \ i = 1, 2, 3, \ldots \tag{2.66}$$

The following theorem indicates the features of the algorithm (2.63)-(2.66).

Theorem 2.4. Under Assumption 2.5, the algorithm (2.63)-(2.66) converges to the exact solution of E with the rate of convergence of $O(\varepsilon^2)$, that is

$$\| E - E^{(i+1)} \| = O(\varepsilon^2) \| E - E^{(i)} \|, \ i = 0, 1, 2, \ldots \tag{2.67}$$

or equivalently

$$\| E - E^{(1)} \| = O(\varepsilon^{2i}) \tag{2.68}$$

Proof: The Jacobian of (2.49)-(2.51), at some $\varepsilon = \varepsilon^*$, is given by

$$J(\varepsilon) = \begin{bmatrix} J_{11}(\varepsilon) & 0 & 0 \\ J_{21}(\varepsilon) & J_{22}(\varepsilon) & J_{23}(\varepsilon) \\ 0 & 0 & J_{33}(\varepsilon) \end{bmatrix} + \begin{bmatrix} O(\varepsilon^2) & O(\varepsilon^2) & 0 \\ O(\varepsilon^2) & O(\varepsilon^2) & O(\varepsilon^2) \\ 0 & O(\varepsilon^2) & O(\varepsilon^2) \end{bmatrix} \tag{2.69}$$

where

$$J_{11}(\varepsilon) = I_{n_1} \oplus D_1^T(\varepsilon) + D_1^T(\varepsilon) \oplus I_{n_1}$$

$$J_{22}(\varepsilon) = I_{n_2} \oplus D_2^T(\varepsilon) + D_1^T(\varepsilon) \oplus I_{n_1} \tag{2.70}$$

$$J_{33}(\varepsilon) = I_{n_2} \oplus D_2^T(\varepsilon) + D_2^T(\varepsilon) \oplus I_{n_2}$$

Since $D_1(\varepsilon)$ and $D_2(\varepsilon)$ are stable matrices (by Assumption 2.5), $J_{ii}(\varepsilon)$, i = 1, 2, 3 are nonsingular and hence the Jacobian will be nonsingular at $\varepsilon = \varepsilon^*$, assuming that ε is sufficiently small. Then by the implicit function theorem, the existence of the unique bouned solution of (2.49)-(2.51) is guaranteed.

In the next step we have to show convergence of the algorithm (2.63)-(2.66) and to give an estimate of the rate of convergence. For i = 0, (2.58) and (2.63) imply

$$(E_1 - E_1^{(1)})D_1 + D_1^T(E_1 - E_1^{(1)}) = \varepsilon^2 g_1(E_1, E_2, \varepsilon^2) \tag{2.71}$$

Since D_1 is stable and E_1 and E_2 are bounded it follows that

$$\| E_1 - E_1^{(1)} \| = O(\varepsilon^2) \tag{2.72}$$

Similarly from (2.59) and (2.64) we have

$$(E_3 - E_3^{(1)})D_2 + D_2^T(E_3 - E_3^{(1)}) = \varepsilon^2 g_3(E_2, E_3, \varepsilon^2) \tag{2.73}$$

and

$$\| E_3 - E_3^{(1)} \| = O(\varepsilon^2) \tag{2.74}$$

The use of the same arguments in (2.60) and (2.65) produces

$$\| E_2 - E_2^{(1)} \| = O(\epsilon^2) \tag{2.75}$$

Continuing the same procedure and by induction we conclude that

$$\| E_1 - E_1^{(i)} \| = O(\epsilon^{2i})$$

$$\| E_2 - E_2^{(i)} \| = O(\epsilon^{2i}) \tag{2.76}$$

$$\| E_3 - E_3^{(i)} \| = O(\epsilon^{2i})$$

with $i = 1, 2, 3 \ldots$, which completes the proof of Theorem 2.4

Numerical examples, that utilize in different set-ups the recursive reduced-order parallel algorithms for the algebraic Lyapunov and Riccati equations of weakly coupled systems, will be presented in next chapters.

2.4 Decoupling Transformation for Weakly Coupled Linear Systems

The linear weakly coupled systems represented by

$$\dot{x} = A_1 x + \epsilon A_2 z + B_1 u_1 + \epsilon B_2 u_2 \tag{2.77}$$

$$\dot{z} = \epsilon A_3 x + A_4 z + \epsilon B_3 u_1 + B_4 u_2 \tag{2.78}$$

where $x \in R^{n_1}$, $z \in R^{n_2}$, $u_i \in R^{m_i}$, $i = 1, 2$, and ϵ is a small parameter has been studied in different set-ups by many researchers (Kokotović, Perkins, Cruz and D'Ans, 1969, Delacour and Fantin, 1978, Petkovski and Rakić, 1979, Mahmoud, 1978, Sezar and Šiljak, 1986, Ishimatsu, Mohri and Takata, 1975, Washburn and Mendel, 1980, Khalil and Kokotović, 1978). The main control equations of weakly coupled linear systems (Riccati type or Lyapunov type) are studied from the power series expansion point of view in (Kokotović, Perkins, Cruz and D'Ans, 1969, Delacour, Darwish and Fantin,

1978, Petkovski and Rakić, 1979, Mahmoud, 1978). A different recursive approach, based on the fixed point iterations, is developed in (Gajić and Rayavarupu, 1989, Petrović and Gajić, 1988, Harkara, Petkovski and Gajić, 1989). In this section we will introduce a new method, that is the nonsingular transformation that completely decouples linear weakly coupled systems (filters or estimators first of all). The main motivation for this transformation is the existence of the corresponding one for the other class of small parameter linear systems - singularly perturbed systems (Kokotović and Khalil, 1986, Chang, 1972). In addition, the proposed transformation completely decouples corresponding Lyapunov differential matrix equation.

Introducing the change of variables

$$x = \eta + \epsilon L z \tag{2.79}$$

the original system (2.77) is transformed into

$$\dot{\eta} = A_{10}\eta + \epsilon F_1(L)z + B_{10}u_1 + \epsilon B_{20}u_2 \tag{2.80}$$

where

$$A_{10} = A_1 - \epsilon^2 L A_3 , \tag{2.81}$$

$$B_{10} = B_1 - \epsilon^2 L B_3 , \quad B_{20} = B_2 - L B_4 \tag{2.82}$$

$$F_1(L) = A_1 L - L A_4 + A_2 - \epsilon^2 L A_3 L \tag{2.83}$$

Assuming that a matrix L can be chosen such that $F_1(L) = 0$, the equation (2.80) will represent a completely independent (decoupled) subsystem

$$\dot{\eta} = A_{10}\eta + B_{10}u_1 + \epsilon B_{20}u_2 \tag{2.84}$$

As a matter of fact, equations (2.78) and (2.84) form a triangular system (after elimination of x from (2.78) by using (2.79))

Introducing a second change of variables as

$$\zeta = z + \epsilon H \eta \tag{2.85}$$

the equation (2.78) becomes

$$\dot{\zeta} = A_{40}\zeta + \varepsilon F_2(H)\eta + \varepsilon B_{30}u_1 + B_{40}u_2 \qquad (2.86)$$

where

$$A_{40} = A_4 + \varepsilon^2 A_3 L \qquad (2.87)$$

$$B_{30} = B_3 + HB_{10}, \quad B_{40} = B_4 + \varepsilon^2 HB_{20} \qquad (2.88)$$

$$F_2(H) = HA_{10} - A_{40}H + A_3 \qquad (2.89)$$

In addition, if matrix H can be chosen such that $F_2(H) = 0$, then we have

$$\dot{\zeta} = A_{40}\zeta + \varepsilon B_{30}u_1 + B_{40}u_2 \qquad (2.90)$$

so that (2.84) and (2.90) represent two completely decoupled linear subsystems. Notice that the weakly coupled structure of the control inputs in (2.77) and (2.78) is preserved in the new coordinates, that is in (2.84) and (2.90). This means that the proposed transformation is applicable to the feedback structure of (2.77) and (2.78) also. Thus, applying the nonsingular transformation

$$\begin{vmatrix} \eta \\ \zeta \end{vmatrix} = \begin{vmatrix} I & -\varepsilon L \\ \varepsilon H & I - \varepsilon^2 HL \end{vmatrix} \begin{vmatrix} x \\ z \end{vmatrix} = T\begin{pmatrix} x \\ z \end{pmatrix} \qquad (2.91)$$

where

$$T^{-1} = \begin{vmatrix} I - \varepsilon^2 LH & \varepsilon L \\ -\varepsilon H & I \end{vmatrix} \qquad (2.92)$$

the linear weakly coupled system (2.77)-(2.78) is completely decoupled and uniquely determined by its subsystems (2.84) and (2.90).

Obviously, the transformation T is uniquely obtained if unique solutions of the following two algebraic equations exist

$$A_1L - LA_4 + A_2 - \varepsilon^2 LA_3L = 0 \tag{2.93}$$

$$H(A_1 - \varepsilon^2 LA_3) - (A_4 + \varepsilon^2 A_3 L)H + A_3 = 0 \tag{2.94}$$

It is important to notice that at $\varepsilon = 0$ we have

$$A_1 L^{(0)} - L^{(0)} A_4 + A_2 = 0 \tag{2.95}$$

$$H^{(0)} A_1 - A_4 H^{(0)} + A_3 = 0 \tag{2.96}$$

so that

$$L = L^{(0)} + O(\varepsilon^2) \tag{2.97}$$

$$H = H^{(0)} + O(\varepsilon^2) \tag{2.98}$$

Equations (2.95) and (2.96) are Sylvester equations and their unique solutions exist if matrices A_1 and A_4 have no eigenvalues in common (Lancaster and Tismenetsky, 1985). Then by the implicit function theorem (Ortega and Rheinboldt, 1970) for a sufficiently small $\varepsilon \in (0, \varepsilon_1]$ there exists a unique solution of weakly nonlinear algebraic equation (2.93). Under the assumption that A_1 and A_4 have no eigenvalues in common and by the fact that the eigenvalues are continuous functions of the matrix elements (Kato, 1980), there exists ε_2 small enough such that for any $\varepsilon \in (0, \varepsilon_2]$ matrices A_{10} and A_{40} will not have eigenvalues in common and thus, the unique solution of (2.94) will exist.

In summary, we have established the following theorem.

Theorem 2.5 Under assumptions that matrices A_1 and A_4 have no eigenvalues in common there exists a small parameter $\varepsilon \in (0, \min(\varepsilon_1, \varepsilon_2)]$ such that the unique solutions of (2.93) and (2.94) exist.

Trajectories of the transformed (decoupled) system are $O(\varepsilon)$ close to the trajectories of the original system. If the coupling parameter ε is extremely small, or if in the design procedure the accuracy of $O(\varepsilon)$ is sufficient, there is no need for the decomposition. However, if $O(\varepsilon)$ is not very small, or if the high accuracy is required, then one needs methods that will produce any desired accuracy, that is the accuracy of $O(\varepsilon^k)$ where $k = 2$,

3, 4, ... Thus, the method proposed in this section is very useful for the intermediate values of ε and for the systems with the high accuracy requirements. In addition, the importance of the proposed transformation is in the design of linear filters and observers - dynamical systems built by the designer. Apparently it is much easier and less expensive to build two dynamical systems of order n_1 and n_2, than one dynamical system of order $n_1 + n_2$.

Note that transformations (2.79) and (2.85) can be used independently to put the system in either lower or upper triangular form. In some applications that might be sufficient.

Numerical solutions for L and H can be obtained by using the fixed point type recursive algorithms similar to those developed in (Gajić, 1986, Gajić and Rayavarupu, 1989, Petrović and Gajić, 1988, Harkara, Petkovski and Gajić, 1989). In the case of the equations (2.93) and (2.94) the corressponding algorithm is given by

$$A_1 L^{(i+1)} - L^{(i+1)} A_4 + A_2 - \varepsilon^2 L^{(i)} A_3 L^{(i)} = 0 \tag{2.99}$$

with $i = 0, 1, 2, ..., N-1$, and $L^{(0)}$ obtained form (2.95)

$$H^{(N)} A_{10}^{(N)} - A_{40}^{(N)} H^{(N)} + A_3 = 0 \tag{2.100}$$

where

$$A_{10}^{(N)} = A_1 - \varepsilon^2 L^{(N)} A_3 , \quad A_{40}^{(N)} = A_4 + \varepsilon^2 A_3 L^{(N)}$$

Using results of (Gajić, 1986, Gajić and Rayavarupu, 1989, Petrović and Gajić, 1988, Harkara, Petkovski and Gajić, 1989), it can be shown similarly that

$$L = L^{(N)} + O(\varepsilon^{2N}) \tag{2.101}$$

and

$$H = H^{(N)} + O(\varepsilon^{2N}) \tag{2.102}$$

hence, the algorithm (2.99) converges with rate of convergence of $O(\varepsilon^2)$.

Example 2.1

In order to demonstrate the efficiency of the proposed algorithm (2.99), we have run a sixth order example. Matrices A_i, i = 1, 2, 3, 4 are chosen randomly (standard deviation = 1 and mean value = 0 for A_1, A_2 and A_3, standard deviation = 2 and mean value = 0 for A_4)

$$A_1 = \begin{bmatrix} -1.720 & -0.999 & -0.592 \\ -1.434 & 0.779 & 0.856 \\ -0.729 & 0.105 & 0.867 \end{bmatrix}, \quad A_2 = \begin{bmatrix} -1.614 & -1.429 & 0.516 \\ 0.225 & 1.928 & 0.310 \\ -0.332 & 0.067 & 0.329 \end{bmatrix}$$

$$A_3 = \begin{bmatrix} -1.398 & 1.039 & 0.557 \\ 1.298 & 1.349 & -0.891 \\ -0.472 & -0.610 & -0.873 \end{bmatrix}, \quad A_4 = \begin{bmatrix} -2.956 & 1.219 & 2.269 \\ -0.038 & -2.240 & 2.296 \\ -0.873 & -2.020 & 2.344 \end{bmatrix}$$

The simulation results for different values of the coupling parameters ε are given in Table 2.2

ε	Number of required iterations such that $\| L - L^{(1)} \|_\infty < 10^{-10}$
0.8	*
0.7	28
0.6	17
0.5	12
0.3	9
0.1	5
0.05	3
0.01	2

* does not converge

Table 2.2 Number of iterations for the fixed point method

Results of Table 2.2 strongly support the necessity for the existence of the recursive scheme for the solution of (2.93), since unless ε is very small, the zeroth and first order approximations are far from the optimal solution.

In Table 2.3, we show the propogation of the error per iteration when ε = 0.1. We can notice that the rate of convergence of the proposed algorithm (2.99) is $O(\varepsilon^2) = O(10^{-2})$

$\varepsilon = 0.1$	$\parallel L - L^{(i)} \parallel_\infty$
i	
0	4.1290×10^{-2}
1	7.4645×10^{-4}
2	1.6401×10^{-5}
3	2.1149×10^{-7}
4	2.0989×10^{-9}

Table 2.3 Error propagation for the fixed point method

The algorithm (2.99) is based on the fixed point iterations, and it will converge as long as the small parameter ε is small enough such that the radius of convergence $\rho(\varepsilon) < 1$ at each iteration.

An alternative way of solving (2.93) is by using the Newton method where the solution of (2.95) plays the role of the initial cor.dition. The Newton method for the similar type of the algebraic equation is presented in (Grodt and Gajić, 1988). The Newton algorithm for (2.93) can be constructed by setting $L^{(i+1)} = L^{(i)} + \Delta L^{(i)}$ and neglecting $O((\Delta L)^2)$ terms. This will produce the Sylvester type equation of the form

$$D_1^{(i)}L^{(i+1)} + L^{(i+1)}D_2^{(i)} = Q^{(i)} \quad , \quad i = 0, 1, 2, \ldots \qquad (2.103)$$

where

$$D_1^{(i)} = \left(A_1 - \varepsilon^2 L^{(i)}A_3\right)$$

$$D_2^{(i)} = -\left(A_4 + \varepsilon^2 A_3 L^{(i)}\right)$$

$$Q^{(i)} = -\left(A_2 + \varepsilon^2 L^{(i)}A_3 L^{(i)}\right)$$

with the initial condition $L^{(0)}$ obtained from (2.95).

The Newton method is demonstrated by solving the same example. For the different values of ε the results are presented in Table 2.4

ε	Number of iterations such that $\parallel L - L^{(1)} \parallel_{\infty} < 10^{-10}$
0.8	5
0.7	5
0.6	4
0.5	4
0.3	3
0.1	2
0.05	2
0.01	1

Table 2.4 Number of iterations for the Newton method

It can be seen, that for this particular example, the Newton method converges much faster than the fixed point iteration algorithm. It is the very well known fact that the Newton method converges quadratically in the neighbourhood of the sought solution and that its main problem is in the choice of the initial guess. For the algebraic equation (2.93) the initial guess is easily obtained with the accuracy of $O(\varepsilon^2)$, and the Newton method, if converges, will produce a sequence $O(\varepsilon^4)$, $O(\varepsilon^8)$, $O(\varepsilon^{16})$, close to the exact solution. However, in some cases the Newton method does not converge at all (bad initial guess) and one needs to have some other efficient techniques available. The fixed point method presented earlier in this section is one of them, since its rate of convergence of $O(\varepsilon^2)$ is remarkable.

Simulation results for all examples solved in this book are obtained by using the software package L-A-S, (West, Bingulac and Perkins, 1985), for computer aided control system design.

In the following we will show that the introduced transformation T completely decouples the corresponding Lyapunov matrix differential equation as well.

Consider the Lyapunov matrix differential equation of weakly coupled systems

$$\dot{P} = A^T P + PA + Q , \quad Q = Q^T , \quad P(t_0) = P_0 \qquad (2.104)$$

where given matrices A and Q are partitioned as

$$A = \begin{vmatrix} A_1 & \varepsilon A_2 \\ \varepsilon A_3 & A_4 \end{vmatrix}, \quad Q = \begin{vmatrix} Q_1 & \varepsilon Q_2 \\ \varepsilon Q_2^T & Q_3 \end{vmatrix}$$

Due to assumed structure for A and Q, the matrix P is properly scaled as (Kokotović, Perkins, Cruz and D'Ans, 1969)

$$P = \begin{vmatrix} P_1 & \varepsilon P_2 \\ \varepsilon P_2^T & P_3 \end{vmatrix} \tag{2.105}$$

Multiplying (2.104) from the left by T^{-T} and from the right hand side by T^{-1} we get

$$T^{-T}\dot{P}T^{-1} = T^{-T}A^TPT^{-1} + T^{-T}PAT^{-1} + T^{-T}QT^{-1} \tag{2.106}$$

which can be written as

$$\dot{K} = a^T K + Ka + q , \quad K(t_0) = K_0 \tag{2.107}$$

where

$$a = TAT^{-1} = \begin{vmatrix} A_{10} & 0 \\ 0 & A_{40} \end{vmatrix} \tag{2.108}$$

$$q = T^{-T}QT^{-1} = \begin{vmatrix} q_1 & \varepsilon q_2 \\ \varepsilon q_2^T & q_3 \end{vmatrix} \tag{2.109}$$

$$K = T^{-T}PT^{-1} = \begin{vmatrix} K_1 & \varepsilon K_2 \\ \varepsilon K_2^T & K_3 \end{vmatrix} , \quad K(t_0) = T^{-T}P_0T^{-1} \tag{2.110}$$

Partitioning (2.107) we can note a completely decoupled form among elements of K

$$\dot{K}_1 = K_1 A_{10} + A_{10}{}^T K_1 + q_1 \qquad\qquad (2.111)$$

$$\dot{K}_2 = K_2 A_{40} + A_{10}{}^T K_2 + q_2 \qquad\qquad (2.112)$$

$$\dot{K}_3 = K_3 A_{40} + A_{40}{}^T K_3 + q_3 \qquad\qquad (2.113)$$

Having obtained K_i's from (2.111)-(2.113), we can get the solution of the Lyapunov differential equation in the original coordinates as

$$P = T^T K T \qquad\qquad (2.114)$$

2.5 Conclusions

In this chapter we have presented the recursive reduced-order parallel algorithms for solving the algebraic Lyapunov and Riccati equations of singularly perturbed and weakly coupled systems. These algorithms are based on the fixed point approach to the small parameter problems, where the small parameter plays the role of the radius of convergence. In many applications, it is shown that these algorithms are computationally very efficient and superior over those based on the power series expansion used so far in the study of the steady state linear estimation and control problems. These results will be exploited later on in subsequent chapters where different types of linear control problems are studied.

In addition powerful transformation that completely decouples weakly coupled linear systems is introduced. Besides its theoretical importance (for example in the stability study and in the study of the variance of linear systems driven by white noise), it can be used in the practical implementation of linear filters and observers (see Chapters 4 and 7).

OUTPUT FEEDBACK CONTROL OF LINEAR SINGULARLY PERTURBED AND WEAKLY COUPLED SYSTEMS

3.1 Introduction

The design of the optimal linear full state regulator requires measurement of all system states. In many practical applications, this is not feasible, due to either the high cost of the state measurements or the inaccessibility for measurement of some of the system states. The standard way to overcome these difficulties is to reconstruct the full state vector from the available measurements by the Luenberger observer, or, if the measurements are noisy, by the Kalman filter. However, these state reconstruction methods will introduce an additional dynamical system. That is why, in the early seventies, increasing attention was given to the problem of designing output constrained regulators where a very limited number of state measurements are available for control implementation (e.g. Levine and Athans, 1970, Levine, Johnson and Athans, 1971, Mendel, 1974, Petkovski and Rakić, 1978). The optimal solution to this control problem is obtained in terms of high order nonlinear matrix algebraic equations. The convergence complexities of the algorithms suggested for the solution of these equations have hindered for quite a long time a wider application of this technique. Recently the convergence problem was solved in (Moerder and Calise, 1985a, Toivonen, 1985). Since that time, the static output feedback control problem has become a very fruitful research area (Makila and Toivonen, 1987).

In this chapter the output feedback control of singularly perturbed and weakly coupled linear systems is studied. The output feedback control problem attracted the attention of the researchers from the field of singular perturbations in the eighties (Calise and Moerder, 1985, Chemouil and Wahdam, 1980, Fossard and Magni, 1980, Khalil, 1981, 1987, Moerder and Calise, 1985b). It is well known that the singularly perturbed systems belong to the class of systems with ill-conditioned dynamics which makes corresponding numerical problems stiff. Thus, in addition to the high order nonlinear matrix algebraic equations, one is faced with the ill-defined numerical problems also.

Motivated by the results of (Gajić, 1986, Gajić and Rayavarupu, 1989, Moerder and Calise, 1985a), we have developed the well-defined recursive numerical technique for the solution of nonlinear algebraic matrix equations associated with the output feedback control problem of linear-quadratic singularly perturbed systems. Moreover, the numerical slow-fast decomposition has been achieved so that only low-order systems are involved in algebraic computations. It is shown that each iteration step of the proposed algorithm improves the accuracy by an order of magnitude, that is, the accuracy of $O(\varepsilon^k)$, where ε is a small perturbation parameter, can be obtained by performing only k iterations. This represents a significant improvement, since all results on the output feedback control problems for the singularly perturbed systems have been obtained so far with an accuracy of $O(\varepsilon)$ only. The real world example, an industrially important reactor, which demonstrates the efficiency of the proposed algorithm and the failure of $O(\varepsilon)$ theory is included in that section.

The output feedback control problem for weakly coupled linear systems is studied in (Petkovski and Rakić, 1979) by using a series expansion approach. This approach is not recursive in application and it is numerically inefficient when a high order of accuracy is required or when the coupling parameter is not very small.

Following the results presented in Chapter 2 a recursive algorithm is developed for solving nonlinear algebraic equations comprising the solution of the optimal static output feedback control problem of linear weakly coupled systems. The numerical decomposition has been achieved so that only low-order systems are involved in algebraic computations. The effectiveness of the proposed reduced-order algorithm and its advantages over the global full-order algorithm are demonstrated on the twelve plate chemical absorption column. Obtained results strongly support the necessity for the existence of the reduced-order numerical techniques for solving corresponding nonlinear algebraic equations. In addition to the reduction in required computations, it would be easier to find a good initial guess and to handle the problem of nonuniqueness of the solution of these nonlinear equations - they represent the necessary conditions only.

3.2 Output Feedback for Singularly Perturbed Linear Systems

Consider the singularly perturbed linear system (Kokotović and Khalil, 1986)

$$\dot{x}_1 = A_1 x_1 + A_2 x_2 + B_1 u , \quad x_1(t_0) = x_{10} \tag{3.1}$$

$$\varepsilon \dot{x}_2 = A_3 x_1 + A_4 x_2 + B_2 u , \quad x_2(t_0) = x_{20} \tag{3.2}$$

$$y = C_1 x_1 + C_2 x_2 \tag{3.3}$$

where $x_1 \in R^{n_1}$ and $x_2 \in R^{n_2}$ are state vectors, $u \in R^m$ is a control input and $y \in R^r$ is a measured output. In the following, A_i, B_j, and C_i, $i = 1, ...4$, $j = 1, 2$ are constant matrices of compatible dimensions, in general they are continuous functions of a small positive parameter ε (Gajić, 1986). With (3.1)-(3.3), consider the performance criterion

$$J = \int_0^\infty \left\{ \begin{bmatrix} x_1 \\ x_2 \end{bmatrix}^T Q \begin{bmatrix} x_1 \\ x_2 \end{bmatrix} + u^T R u \right\} dt \tag{3.4}$$

with positive definite R and positive semidefinite Q, which has to be minimized. In addition, the control input $u(t)$ is constrained to

$$u(t) = - Fy(t) \tag{3.5}$$

The optimal constant output feedback gain F is given by (Levine and Athans, 1970)

$$F = R^{-1}B^TKLC^T(CLC^T)^{-1} \tag{3.6}$$

where matrices K and L satisfy high order nonlinear coupled algebraic equations

$$(A - BFC)L + L(A - BFC)^T + x_0x_0^T = 0 \tag{3.7}$$

$$(A - BFC)^TK + K(A - BFC) + Q + C^TF^TRFC = 0 \tag{3.8}$$

and newly defined matrices as

$$A = \begin{bmatrix} A_1 & A_2 \\ \dfrac{A_3}{\varepsilon} & \dfrac{A_4}{\varepsilon} \end{bmatrix}, \quad B = \begin{bmatrix} B_1 \\ \dfrac{B_2}{\varepsilon} \end{bmatrix}, \quad C = \begin{bmatrix} C_1 & C_2 \end{bmatrix}, \quad x_0 = \begin{bmatrix} x_{10} \\ x_{20} \end{bmatrix} \tag{3.9}$$

Compatible to the nature of their solution, matrices K and L are partitioned as follows

$$K = \begin{bmatrix} K_1 & \varepsilon K_2 \\ \varepsilon K_2^T & \varepsilon K_3 \end{bmatrix}, \quad L = \begin{bmatrix} L_1 & L_2 \\ L_2^T & L_3 \end{bmatrix} \tag{3.10}$$

In a recent paper (Moerder and Calise, 1985a), it is shown that the algorithm proposed for the numerical solution of (3.6)-(3.8), defined by

Choose $F^{(0)}$ such that $A - BF^{(0)}C$ is a stable matrix $\qquad$ (3.11)

$$(A - BF^{(i)}C)L^{(i+1)} + L^{(i+1)}(A - BF^{(i)}C)^T + x_0x_0^T = 0 \tag{3.12}$$

$$(A - BF^{(i)}C)^TK^{(i+1)} + K^{(i+1)}(A - BF^{(i)}C) + Q + C^TF^{(i)T}RF^{(i)}C = 0 \tag{3.13}$$

$$F^{(i+1)} = R^{-1}B^TK^{(i+1)}L^{(i+1)}C^T(CL^{(i+1)}C^T)^{-1} \tag{3.14}$$

with $i = 1, 2, ...$, converges to a local minimum under the nonrestrictive assumption. As a matter of fact, the updated value for F is defined in (Moerder and Calise, 1985a) as

$$F_N^{(i+1)} = F^{(i)} + \alpha\left(F^{(i+1)} - F^{(i)}\right) \tag{3.15}$$

where $\alpha \in (0, 1]$ is chosen at each iteration to ensure that the minimum is not overshoot, that is

$$J_{i+1} = tr\left\{K^{(i+1)}x_0x_0^T\right\} < J_i = tr\left\{K^{(i)}x_0x_0^T\right\} \tag{3.16}$$

It has been customary in the control literature on the output feedback to assume that the initial conditions are uniformly distributed on the unit sphere, that is,

$$E\left\{x_0x_0^T\right\} = I_{(n_1+n_2)} \tag{3.17}$$

Applying the slow-fast decomposition transform of Chang (Kokotović and Khalil, 1986) to problem (3.1)-(3.5) and finding the optimal gains for the slow and fast subsystem is possible for the accuracy of $O(\varepsilon)$ only (Calise and Moerder, 1985; Moerder and Calise, 1985b). It represents a well-posed problem, but there is no way to improve the approximation to any desired order of accuracy, that is $O(\varepsilon^k)$. In this section, we will achieve that goal through the numerical slow-fast decomposition of the algebraic equations (3.11)-(3.15)

In order to simplify derivations, we introduce a notation

$$A - BFC = \begin{bmatrix} D_1 & D_2 \\ \dfrac{D_3}{\varepsilon} & \dfrac{D_4}{\varepsilon} \end{bmatrix} = \begin{bmatrix} A_1 - B_1FC_1 & A_2 - B_1FC_2 \\ \dfrac{A_3 - B_2FC_1}{\varepsilon} & \dfrac{A_4 - B_2FC_2}{\varepsilon} \end{bmatrix} \tag{3.18}$$

$$Q + C^TF^TRFC = \begin{bmatrix} q_1 & q_2 \\ q_3 & q_4 \end{bmatrix} = \begin{bmatrix} Q_1 + C_1^TF^TRFC_1 & Q_2 + C_1^TF^TRFC_2 \\ Q_2^T + C_2^TF^TRFC_1 & Q_3 + C_2^TF^TRFC_2 \end{bmatrix} \tag{3.19}$$

with obvious definitions for D_i's and q_i's, $i = 1, 2, 3, 4$.

Partitioning (3.12)–(3.13) compatible to (3.9)-(3.10) and using (3.17)-(3.19) will produce the following set of equations

$$D_1^{(i)}L_1^{(i+1)} + L_1^{(i+1)}D_1^{(i)T} + D_2^{(i)}L_2^{(i+1)T} + L_2^{(i+1)}D_2^{(i)T} + I = 0 \qquad (3.20)$$

$$L_2^{(i+1)}D_4^{(i)T} + \varepsilon D_1^{(i)}L_2^{(i+1)} + L_1^{(i+1)}D_3^{(i)T} + \varepsilon D_2^{(i)}L_3^{(i+1)} = 0 \qquad (3.21)$$

$$L_3^{(i+1)}D_4^{(i)T} + D_4^{(i)}L_3^{(i+1)} + D_3^{(i)}L_2^{(i+1)} + L_2^{(i+1)T}D_3^{(i)T} + \varepsilon I = 0 \qquad (3.22)$$

and

$$D_1^{(i)T}K_1^{(i+1)} + K_1^{(i+1)}D_1^{(i)} + D_3^{(i)T}K_2^{(i+1)T} + K_2^{(i+1)}D_3^{(i)} + q_1^{(i)} = 0 \quad (3.23)$$

$$K_2^{(i+1)}D_4^{(i)} + \varepsilon D_1^{(i)T}K_2^{(i+1)} + D_3^{(i)T}K_3^{(i+1)} + K_1^{(i+1)}D_2^{(i)} + q_2^{(i)} = 0 \quad (3.24)$$

$$K_3^{(i+1)}D_4^{(i)} + D_4^{(i)T}K_3^{(i+1)} + \varepsilon D_2^{(i)T}K_2^{(i+1)} + \varepsilon K_2^{(i+1)T}D_2^{(i)} + q_3^{(i)} = 0$$
$$(3.25)$$

where

$$D_1^{(i)} = A_1 - B_1 F^{(i)} C_1 , \qquad D_2^{(i)} = A_2 - B_1 F^{(i)} C_2$$

$$D_3^{(i)} = A_3 - B_2 F^{(i)} C_1 , \qquad D_4^{(i)} = A_4 - B_2 F^{(i)} C_2$$

and

$$q_1^{(i)} = Q_1 + C_1^T F^{(i)T} R F^{(i)} C_1$$

$$q_2^{(i)} = Q_2 + C_1^T F^{(i)T} R F^{(i)} C_2$$

$$q_3^{(i)} = Q_3 + C_2^T F^{(i)T} R F^{(i)} C_2 \quad , \ i = 0, 1, 2, 3, \ldots.$$

Since the matrix $A-BF^{(i)}C$ has n_1 slow eigenvalues of $O(1)$ and n_2 fast eigenvalues of $O(\frac{1}{\varepsilon})$, then $\det(A-BF^{(i)}C)$ is of $O(1/\varepsilon^{n_2})$, which makes (3.12) and (3.13) numerically ill-defined. However, the partitioned forms of (3.12) and (3.13) given by (3.20)–(3.22) and (3.23)-(3.25), obtained after multiplying equations for $L_2(K_2)$ and $L_3(K_3)$ by ε, comprise the well-defined numerical problems, but there are no available methods for their solution. In what follows we will derive the efficient numerical scheme for solving (3.20)-(3.22) and (3.23)-(3.25). Even more, the slow-fast decomposition will be achieved, and the required solutions will be obtained in terms of low-order problems of dimensions n_1 and n_2 - the original problems (3.20)-(3.22) and (3.23)-(3.25) are of dimensions $n_1 + n_2$.

The equations (3.23)-(3.25) form a standard Lyapunov equation of singularly perturbed linear systems. It is a special case of the more general Lyapunov equation studied in Chapter 2. Its zeroth-order solution is obtained by setting $\varepsilon = 0$ in (3.23)-(3.25), which after some algebra produces

$$\underline{K}_1^{(i+1)}D_0^{(i)} + D_0^{(i)T}\underline{K}_1^{(i+1)} + G_0^{(i)T}G_0^{(i)} = 0 \tag{3.26}$$

$$\underline{K}_3^{(i+1)}D_4^{(i)} + D_4^{(i)T}\underline{K}_3^{(i+1)} + q_3^{(i)} = 0 \tag{3.27}$$

$$\underline{K}_2^{(i+1)} = - \left(\underline{K}_1^{(i+1)}D_2^{(i)} + D_3^{(i)T}\underline{K}_3^{(i+1)} + q_2^{(i)}\right)D_4^{(i)-1} \tag{3.28}$$

where

$$D_0^{(i)} = D_1^{(i)} - D_2^{(i)}D_4^{(i)-1}D_3^{(i)}$$

$$G_0^{(i)} = G_1^{(i)} - G_3^{(i)}D_4^{(i)-1}D_3^{(i)} \ , \ G_p^{(i)} = \sqrt{q_p^{(i)}} \ , \ p = 1, 3$$

Note that there is no need to calculate square root of $q_p^{(i)}$'s. The expression for $G_0^{(i)}$ is used in (3.26) only to simplify notation, but not for real calculations since

$$q_1^{(i)} = G_1^{(i)T}G_1^{(i)} \ , \ q_2^{(i)} = G_1^{(i)T}G_3^{(i)} \ , \ \text{and } q_3^{(i)} = G_3^{(i)T}G_3^{(i)}.$$

The zeroth-order solution

$$\underline{K}^{(i+1)} = \begin{bmatrix} \underline{K}_1^{(i+1)} & \varepsilon\underline{K}_2^{(i+1)} \\ \varepsilon\underline{K}_2^{(i+1)T} & \varepsilon\underline{K}_3^{(i+1)} \end{bmatrix} \tag{3.29}$$

is $O(\varepsilon)$ close to the required one $K^{(i+1)}$. We can relate them trough the error term E

$$\epsilon E = K^{(i+1)} - \underline{K}^{(i+1)} \tag{3.30}$$

or by using a compatible partition:

$$\begin{bmatrix} \epsilon E_1 & \epsilon^2 E_2 \\ \epsilon^2 E_2^T & \epsilon^2 E_3 \end{bmatrix} = \begin{bmatrix} K_1^{(i+1)} - \underline{K}_1^{(i+1)} & \epsilon\left(K_2^{(i+1)} - \underline{K}_2^{(i+1)}\right) \\ \epsilon\left(K_2^{(i+1)} - \underline{K}_2^{(i+1)}\right)^T & \epsilon\left(K_3^{(i+1)} - \underline{K}_3^{(i+1)}\right) \end{bmatrix} \tag{3.31}$$

Clearly the $O(\epsilon^k)$ approximation of E will produce the $O(\epsilon^{k+1})$ approximation of the sought solution $K^{(i+1)}$, which is why we are interested in finding a convenient form for the error equation and an appropriate algorithm for its solution. It is shown in Section 2.2.1 that the error equation is given by

$$D_1^{(i)T} E_1 + E_1 D_1^{(i)} + D_3^{(i)T} E_2^T + E_2 D_3^{(i)} = 0 \tag{3.32}$$

$$E_2 D_4^{(i)} + \epsilon D_1^{(i)T} E_2 + D_1^{(i)T} \underline{K}_2^{(i+1)} + D_3^{(i)T} E_3 = 0 \tag{3.33}$$

$$E_3 D_4^{(i)} + D_4^{(i)T} E_1^{(i+1)} + D_2^{(i)T} \underline{K}_2^{(i+1)} + \underline{K}_2^{(i+1)T} D_2^{(i)}$$
$$+ \epsilon\left(D_2^{(i)T} E_2 + E_2^T D_2^{(i)}\right) = 0 \tag{3.34}$$

and that the following algorithm

$$D_0^{(i)T} E_1^{(j+1)} + E_1^{(j+1)} D_0^{(i)} = D_0^{(i)T}\left(\underline{K}_2^{(i+1)} + \epsilon E_2^{(j)}\right) D_4^{(i)-1} D_3^{(i)} +$$
$$D_3^{(i)T} D_4^{(i)-T}\left(\underline{K}_2^{(i+1)} + \epsilon E_2^{(j)}\right)^T D_0^{(i)} \tag{3.35}$$

$$D_4^{(i)T} E_3^{(j+1)} + E_3^{(j+1)} D_4^{(i)} = -D_2^{(i)T}\left(\underline{K}_2^{(i+1)} + \epsilon E_2^{(j)}\right) -$$
$$\left(\underline{K}_2^{(i+1)} + \epsilon E_2^{(j)}\right)^T D_2^{(i)} \tag{3.36}$$

$$E_2^{(j+1)} = - D_3^{(i)T} E_3^{(j+1)} + E_1^{(j+1)} D_2^{(i)} + D_1^{(i)T}\left(\underline{K}_2^{(i+1)} + \epsilon E_2^{(j)}\right) D_4^{(i)-1} \tag{3.37}$$

$$j = 1, 2, 3, \ldots$$

with initial conditions chosen as $E_1^{(0)} = 0$, $E_2^{(0)} = 0$, and $E_3^{(0)} = 0$, converges to required solution E with the rate of convergence of $O(\varepsilon)$, that is,

$$\| E - E^{(j)} \| = O(\varepsilon^j) \, , \, j = 1, \, 2, \, 3, \, ... \tag{3.38}$$

That implies

$$\| K^{(i+1)} - \left(\underline{K}^{(i+1)} + \varepsilon E^{(j)} \right) \| = O(\varepsilon^j) \, , \, j = 1, \, 2, \, 3, \, ... \tag{3.39}$$

Note that the complete slow-fast decomposition is achieved, that is, the solution of the Lyapunov equations (3.23)-(3.25) of order $n_1 + n_2$ is obtained in terms of two low-order Lyapunov equations, the slow one (3.35) of order n_1, and the fast one (3.36) of order n_2.

The equations (3.20)-(3.22) do not represent a standard Lyapunov equations of singularly perturbed systems due to the fact that the initial conditions satisfy (3.17). In the following, we apply the methodology of (Gajić, 1986; Gajić and Rayavarupu, 1989) to (3.20)-(3.22) subject to (3.17), and derive the recursive algorithm for its solution in terms of the reduced-order problems.

Setting $\varepsilon = 0$ in (3.20)-(3.22) will produce, after some algebra, the zeroth-order approximation of (3.20)-(3.23) as

$$\underline{L}_1^{(i+1)}D_0^{(i)T} + D_0^{(i)}\underline{L}_1^{(i+1)} + I = 0 \tag{3.40}$$

$$\underline{L}_2^{(i+1)} = -\underline{L}_1^{(i+1)}D_3^{(i)T}D_4^{(i)-T} \tag{3.41}$$

$$\underline{L}_3^{(i+1)}D_4^{(i)T} + D_4^{(i)}\underline{L}_3^{(i+1)} + \underline{L}_2^{(i+1)T}D_3^{(i)T} + D_3^{(i)}\underline{L}_2^{(i+1)} = 0 \tag{3.42}$$

Even though the complete slow-fast decomposition is not achieved (in contrary to (3.26)-(3.28)), these equations can be solved in terms of reduced-order problems in a sequential manner, namely, first solve (3.40), then (3.41) and finally solve (3.42).

Defining the error as

$$L^{(i+1)} - \underline{L}^{(i+1)} = \varepsilon M = \varepsilon \begin{bmatrix} M_1 & M_2 \\ M_2^T & M_3 \end{bmatrix} =$$

$$= \begin{bmatrix} L_1^{(i+1)} - \underline{L}_1^{(i+1)} & L_2^{(i+1)} - \underline{L}_2^{(i+1)} \\ (L_2^{(i+1)} - \underline{L}_2^{(i+1)})^T & L_3^{(i+1)} - \underline{L}_3^{(i+1)} \end{bmatrix} \quad (3.43)$$

and subtracting (3.40)-(3.42) from (3.20)-(3.22), we get the error equation as

$$M_1 D_1^{(i)T} + D_1^{(i)}M_1 + D_2^{(i)}M_2^T + M_2 D_2^{(i)T} = 0 \qquad (3.44)$$

$$M_2 D_4^{(i)T} + \varepsilon D_1^{(i)}M_2 + D_1^{(i)}\underline{L}_2^{(i+1)} + D_2^{(i)}\underline{L}_3^{(i+1)}$$
$$+ \varepsilon D_2^{(i)}M_3 + M_1 D_3^{(i)T} = 0 \qquad (3.45)$$

$$M_3 D_4^{(i)T} + D_4^{(i)}M_3 + D_3^{(i)}M_2 + M_2^T D_3^{(i)T} + I = 0 \qquad (3.46)$$

Note that (3.45) is a weakly linear Lyapunov equation. At this point, we will ignore that fact and solve it with respect to M_2 as follows:

$$M_2 = -[D_1^{(i)}(\underline{L}_2^{(i+1)} + \varepsilon M_2) + D_2^{(i)}(\underline{L}_3^{(i+1)} + \varepsilon M_3) + M_1 D_3^{(i)T}]D_4^{(i)-T} \quad (3.47)$$

Using (3.47) in (3.44) yields

$$M_1 D_0^{(i)T} + D_0^{(i)}M_1 - D_2^{(i)}D_4^{(i)-1}H_2 - H_2 D_4^{(i)-T}D_2^{(i)T} = 0 \qquad (3.48)$$

where

$$H_2 = D_1^{(i)}(\underline{L}_2^{(i+1)} + \varepsilon M_2) + D_2^{(i)}(\underline{L}_3^{(i+1)} + \varepsilon M_3) =$$
$$D_1^{(i)}L_2^{(i+1)} + D_2^{(i)}L_3^{(i+1)} \qquad (3.49)$$

Thus, the weakly coupled and hierarchical structure of (3.44)-(3.46) can be exploited by proposing the following recursive scheme, which leads to the two low-order completely decoupled Lyapunov equations

$$M_1^{(j+1)}D_0^{(i)T} + D_0^{(i)}M_1^{(j+1)} - D_2^{(i)}D_4^{(i)-1}H_2^{(i)T} - H_2^{(i)}D_4^{(i)-T}D_2^{(i)T} = 0$$

$$(3.50)$$

$$M_2^{(j+1)} = -\left[H^{(j)} + M_1^{(j+1)}D_3^{(i)T}\right]D_4^{(i)-T}$$

$$(3.51)$$

$$M_3^{(j+1)}D_4^{(i)T} + D_4^{(i)}M_3^{(j+1)} + D_3^{(i)}M_2^{(j+1)} + M_2^{(j+1)T}D_3^{(i)T} + I = 0 \qquad (3.52)$$

where

$$H^{(j)} = D_1^{(i)}\left(L_2^{(i+1)} + \varepsilon M_2^{(j)}\right) + D_2^{(i)}\left(L_3^{(i+1)} + \varepsilon M_3^{(j)}\right)$$

$$(3.53)$$

with $j = 0, 1, 2, 3, \ldots$, with initial conditions chosen as $M_1^{(0)} = 0$, $M_2^{(0)} = 0$, and $M_3^{(0)} = 0$.

The following theorem summarizes the features of the proposed scheme, (Gajić, Petkovski, Harkara, 1989).

Theorem 3.1 The algorithm (3.50)-(3.53) converges, for sufficiently small values of ε, to the exact solution of the error terms, and thus to the solution $L^{(i+1)}$, with the rate of convergence of $O(\varepsilon)$, that is,

$$\left\|M_k - M_k^{(j)}\right\| = O(\varepsilon^j) , \quad k = 1, 2, 3$$

3.3 Case Study: Fluid Catalytic Cracker

In order to demonstrate the efficiency of the proposed algorithm and the failure of the $O(\varepsilon)$ theory, we have run a fifth order real world example, an industrially important reactor (Arkin and Ramakrishnan, 1983). Matrices A, B, C, Q, and R are given in (Arkin and Ramakrishnan, 1983). The eigenvalues of the matrix A are -2.8, -7.7, -74, -82, -129. Thus, we have two slow and three fast variables. The small parameter ε is chosen as $\varepsilon = 0.1$, which is roughly the ratio of 7.7 and 74.

The theory of singularly perturbed optimal output feedback problems is derived so far for the $O(\varepsilon)$ approximation. Using the $O(\varepsilon)$ approximation of the equations comprising the solution of the optimal output feedback, namely of (3.26)-(3.28) and (3.40)-(3.42), will fail to produce the desired approximation for this example. Even more, the algorithm does not converge to the near optimum solution for the extremely small values of the parameter α such as 0.001. The cause of the trouble is the inversion of the quantity CLC^T. Its determinant for the optimal value of L is very small, that is, 0.9736×10^{-4}, and thus, this problem is very sensitive to $O(\varepsilon)$ perturbations, which can be seen from Table 3.1.

α	$\det(CL^{(1)}C^T)$ $j = 6$	$\det(CL^{(1)}C^T)$ $j = 1$	$\det(C\underline{L}^{(0)}C^T)$	$\det(C\underline{L}^{(1)}C^T)$
0.5	0.86846×10^{-4}	0.14432×10^{-3}	0.38943×10^{-6}	0.24392×10^{-10}
0.1	0.12749×10^{-3}	"	"	0.57491×10^{-9}
0.01	0.14244×10^{-3}	"	"	0.31742×10^{-7}
0.001	0.14413×10^{-3}	"	"	0.24904×10^{-6}

Table 3.1 Determinat of CLC^T

The results from Table 3.1 strongly support the necessity for the existence of the recursive schemes which can produce any desired accuracy, that is, the development of the $O(\varepsilon^k)$ theory.

In Table 3.2 we have presented results for the criterion $J_{opt}^{(1)}$ and the gain error for the global algorithm (Moerder and Calise, 1985a), and the corresponding quantities for the proposed reduced-order recursive algorithm. The initial value for the gain $F^{(0)}$ is obtained from (Petkovski and Rakić, 1978). It can be seen that the initial guess is quite good, but the global algorithm converges very slowly to the optimal solution. As far as the criterion is concerned it takes 28 iterations to achieve an accuracy of up to five decimal digits, where J_{opt} = 0.28573. On the other hand, the trajectories of the approximate system after 30 iterations are still far apart from the optimal trajectories since the approximate gain is only $O(10^{-2})$ close to the optimal one. Thus, this algorithm demands a lot of iterations in order to achieve high accuracy. This fact justifies even more the necessity for the existence of algorithms which will reduce computational requirements. In the proposed algorithm, only low-order Lyapunov equations are involved in algebraic computations. Even more, at the very beginning, they can be solved with reduced accuracy (j = 1 or 2), and once we approach the optimum, the accuracy can be increased to the desired one. The third column of Table 3.2 is obtained for j = 2, for $i \le 16$, and j = 6 for i > 16. The second and fifth columns of Table 3.2 are obtained for j = 6 for all i's. The parameter α is chosen as $\alpha = 0.5$ since the global algorithm does not converge for $\alpha \geq 0.6$.

$\varepsilon = 0.1$ $\alpha = 0.5$ i	$J_{opt}^{(i)}$	$J_{app}^{(i)}$ j=6	$J_{app}^{(i)}$ j=2, i≤16 j=6, i>16	$\|F_{opt}^{(i)} - F_{opt}\|_\infty$	$\|F_{app}^{(i)} - F_{opt}\|_\infty$ j=6
1	0.30487	0.30488	0.30427	2.1520	2.1480
2	0.28733	0.28738	0.28879	0.1635	0.1684
4	0.28615	0.28619	0.28745	0.1296	0.1328
6	0.28595	0.28599	0.28710	0.1093	0.1120
8	0.28588	0.28591	0.28691	0.0913	0.0936
10	0.28583	0.28586	0.28676	0.0764	0.0783
12	0.28580	0.28583	0.28664	0.0638	0.0654
14	0.28578	0.28580	0.28654	0.0533	0.0550
16	0.28577	0.28578	0.28646	0.0446	0.0456
18	0.28575	0.28577	0.28584	0.0373	0.0380
20	0.28575	0.28576	0.28581	0.0311	0.0317
22	0.28574	0.28576	0.28579	0.0260	0.0256
24	0.28574	0.28575	0.28577	0.0217	0.0219
26	0.28574	0.28575	0.28577	0.0181	0.0181
28	0.28573	0.28575	0.28576	0.0150	0.0149
30	0.28573	0.28575	0.28575	0.0125	0.0122

Table 3.2 Optimal and approximate criteria

3.4 Output Feedback for Linear Weakly Coupled Systems

Consider the weakly coupled linear system (Petkovski and Rakić, 1979)

$$\dot{x}_1 = A_1 x_1 + \varepsilon A_2 x_2 + B_1 u_1 + \varepsilon B_2 u_2 , \qquad x_1(t_0) = x_{10} \tag{3.54}$$

$$\dot{x}_2 = \varepsilon A_3 x_1 + A_4 x_2 + \varepsilon B_3 u_1 + B_4 u_2 , \qquad x_2(t_0) = x_{20} \tag{3.55}$$

$$y = \begin{bmatrix} C_1 & \varepsilon C_2 \\ \varepsilon C_3 & C_4 \end{bmatrix} \begin{bmatrix} x_1 \\ x_2 \end{bmatrix} = \begin{bmatrix} y_1 \\ y_2 \end{bmatrix} \tag{3.56}$$

where $x_1 \in R^{n_1}$ and $x_2 \in R^{n_2}$ are state vectors, $u_i \in R^{m_i}$, $i = 1, 2$ are control inputs and $y_i \in R^{r_i}$, $i = 1, 2$ are measured outputs. In the following, A_i, B_i and C_i, $i = 1, ..., 4$, are constant matrices of compatible dimensions, in general they are continuous functions of a small parameter ε (Petrović and Gajić, 1988). With (3.54)-(3.56), consider the performance criterion,

$$J = \int_0^{\infty} \left\{ \begin{bmatrix} x_1 \\ x_2 \end{bmatrix}^T Q \begin{bmatrix} x_1 \\ x_2 \end{bmatrix} + u^T R u \right\} dt \ . \tag{3.57}$$

with

$$Q = \begin{bmatrix} Q_1 & \varepsilon Q_2 \\ \varepsilon Q_2^T & Q_4 \end{bmatrix} \quad \text{and} \quad R = \begin{bmatrix} R_1 & 0 \\ 0 & R_4 \end{bmatrix}$$

with positive-definite R and positive-semi-definite Q, which has to be minimized.

In addition, the control input u(t) is constrained to be a direct feedback from the output y(t), eqn. (3.5).

The optimal constant output feedback gain F is given by (Levine and Athans, 1970)

$$F = R^{-1} B^T K L C^T (C L C^T)^{-1} = \begin{bmatrix} F_1 & \varepsilon F_2 \\ \varepsilon F_3 & F_4 \end{bmatrix} \tag{3.58}$$

where matrices K and L satisfy high-order nonlinear coupled algebraic equations (3.7) and (3.8) and newly defined matrices A, B, and C are

$$A = \begin{bmatrix} A_1 & \varepsilon A_2 \\ \varepsilon A_3 & A_4 \end{bmatrix} \quad B = \begin{bmatrix} B_1 & \varepsilon B_2 \\ \varepsilon B_3 & B_4 \end{bmatrix} \quad C = \begin{bmatrix} C_1 & \varepsilon C_2 \\ \varepsilon C_3 & C_4 \end{bmatrix} \tag{3.59}$$

Compatible with the nature of their solution, matrices K and L are partitioned as

$$K = \begin{bmatrix} K_1 & \varepsilon K_2 \\ \varepsilon K_2{}^T & K_3 \end{bmatrix} \quad \text{and} \quad L = \begin{bmatrix} L_1 & \varepsilon L_2 \\ \varepsilon L_2{}^T & L_3 \end{bmatrix} \tag{3.60}$$

The algorithm given by equations (3.11)-(3.14), described in Section 3.2, can be applied for the numerical solution of the corresponding nonlinear matrix equations (3.6)-(3.8) with A, B and C now defined by (3.59).

In order to simplify derivations, the following notation is introduced:

$$\begin{bmatrix} A - BFC \end{bmatrix} = \begin{bmatrix} D_1 & \varepsilon D_2 \\ \varepsilon D_3 & D_4 \end{bmatrix} = \begin{bmatrix} A_1 - B_1 F_1 C_1 & \varepsilon(A_2 - B_1 F_2 C_4) \\ \varepsilon(A_3 - B_4 F_3 C_1) & A_4 - B_4 F_4 C_4 \end{bmatrix} \tag{3.61}$$

Also

$$\begin{bmatrix} Q + C^T F^T RFC \end{bmatrix} = \begin{bmatrix} q_1 & \varepsilon q_2 \\ \varepsilon q_2{}^T & q_3 \end{bmatrix} \tag{3.62}$$

where

$$q_1 = Q_1 + C_1{}^T F_1{}^T R_1 F_1 C_1 + C_1{}^T F_3{}^T R_4 F_3 C_1$$

$$q_2 = Q_2 + C_1{}^T F_1{}^T R_1 F_2 C_4 + C_1{}^T F_3{}^T R_4 F_4 C_4$$

$$q_3 = Q_4 + C_4{}^T F_2{}^T R_1 F_2 C_4 + C_4{}^T F_4{}^T R_4 F_4 C_4$$

In addition, without loss of generality, it has been assumed that matrices B_2, B_3, C_2 and C_3 are zero.

Partitioning (3.12) and (3.13) compatible to (3.59) and (3.60) and using (3.61) and (3.62), we get the following set of equations

$$D_1^{(i)}L_1^{(i+1)} + L_1^{(i+1)}D_1^{(i)T} + \varepsilon^2(D_2^{(i)}L_2^{(i+1)T} + L_2^{(i+1)}D_2^{(i)T}) + I = 0 \quad (3.63)$$

$$L_2^{(i+1)}D_4^{(i)T} + D_1^{(i)}L_2^{(i+1)} + L_1^{(i+1)}D_3^{(i)T} + D_2^{(i)}L_3^{(i+1)} = 0 \quad (3.64)$$

$$L_3^{(i+1)}D_4^{(i)T} + D_4^{(i)}L_3^{(i+1)} + \varepsilon^2(D_3^{(i)}L_2^{(i+1)} + L_2^{(i+1)T}D_3^{(i)T}) + I = 0$$
$$(3.65)$$

and

$$D_1^{(i)T}K_1^{(i+1)} + K_1^{(i+1)}D_1^{(i)} + \varepsilon^2(D_3^{(i)T}K_2^{(i+1)T} + K_2^{(i+1)}D_3^{(i)}) + q_1^{(i)} = 0$$
$$(3.66)$$

$$K_2^{(i+1)}D_4^{(i)} + D_1^{(i)T}K_2^{(i+1)} + D_3^{(i)T}K_3^{(i+1)} + K_1^{(i+1)}D_2^{(i)} + q_2^{(i)} = 0 \quad (3.67)$$

$$K_3^{(i+1)}D_4^{(i)} + D_4^{(i)T}K_3^{(i+1)} + q_3^{(i)} +$$
$$\varepsilon^2(D_2^{(i)T}K_2^{(i+1)} + K_2^{(i+1)T}D_2^{(i)}) = 0 \quad (3.68)$$

where

$$D_1^{(i)} = A_1 - B_1F_1^{(i)}C_1, \qquad D_2^{(i)} = A_2 - B_1F_2^{(i)}C_4$$

$$D_3^{(i)} = A_3 - B_3F_3^{(i)}C_1, \qquad D_4^{(i)} = A_4 - B_4F_4^{(i)}C_2$$

and

$$q_1^{(i)} = Q_1 + C_1^T F_1^{(i)T}R_1F_1^{(i)}C_1 + C_1^TF_3^{(i)T}R_4F_3^{(i)}C_1$$

$$q_2^{(i)} = Q_2 + C_1^TF_1^{(i)T}R_1F_2^{(i)}C_4 + C_1^TF_3^{(i)T}R_4F_4^{(i)}C_4$$

$$q_3^{(i)} = Q_4 + C_4^TF_2^{(i)T}R_1F_2^{(i)}C_4 + C_4^TF_4^{(i)T}R_4F_4^{(i)}C_4$$

with i = 0, 1, 2, 3, ...

Equations (3.63)-(3.68) were studied in (Petkovski and Rakić, 1979) by using a series expansion method. That approach is not recursive in application and it is numerically justified in (Petkovski and Rakić, 1979) for $O(\varepsilon)$ accuracy only. In this section we intend to develop a recursive scheme which will extend efficiently the main results of (Petkovski and

Rakić, 1979) to any arbitrary order of accuracy, namely $O(\varepsilon^{2k})$, where k represents the number of required iterations of the proposed recursive scheme.

Note that all of matrices D_1, K_1, L_1, and q_1 are functions of a small parameter ε. However, dependence on ε is suppressed in order to simplify notation.

Stability of the matrix $D^{(i)}(\varepsilon)$ given by

$$D^{(i)}(\varepsilon) = \begin{bmatrix} D_1^{(i)}(\varepsilon) & \varepsilon D_2^{(i)}(\varepsilon) \\ \varepsilon D_3^{(i)}(\varepsilon) & D_4^{(i)}(\varepsilon) \end{bmatrix}$$

is guaranteed by (Moerder and Calise, 1985a). Due to block diagonal dominance of $D^{(i)}(\varepsilon)$, matrices $D_1^{(i)}(\varepsilon)$ and $D_4^{(i)}(\varepsilon)$ are stable for a sufficiently small values of ε for $\forall i$.

In what follows equations (3.63)-(3.68) will be numerically solved in terms of the reduced-order Lyapunov and Sylvester equations by using results from Chapter 2.

Notice that the equations (3.63)-(3.68) represent standard Lyapunov equations of weakly coupled linear systems. An efficient recursive algorithm for their numerical solution, with $O(\varepsilon^2)$ rate of convergence, has been derived in Section 2.3.1. The zeroth-order solutions of these equations are obtained by setting $\varepsilon = 0$ in (3.63)-(3.68)

$$D_1^{(i)} L_1^{(i+1)} + L_1^{(i+1)} D_1^{(i)T} + I = 0 \tag{3.69}$$

$$L_2^{(i+1)} D_4^{(i)T} + D_1^{(i)} L_2^{(i+1)} + L_1^{(i+1)} D_3^{(i)T} + D_2^{(i)} L_3^{(i+1)} = 0 \tag{3.70}$$

$$L_3^{(i+1)} D_4^{(i)T} + D_4^{(i)} L_3^{(i+1)} + I = 0 \tag{3.71}$$

and

$$D_1^{(i)^T} \underline{K}_1^{(i+1)} + \underline{K}_1^{(i+1)} D_1^{(i)} + q_1^{(i)} = 0 \tag{3.72}$$

$$\underline{K}_2^{(i+1)} D_4^{(i)} + D_1^{(i)^T} \underline{K}_2^{(i+1)} + D_3^{(i)^T} \underline{K}_3^{(i+1)} + \underline{K}_1^{(i+1)} D_2^{(i)} + q_2^{(i)} = 0 \tag{3.73}$$

$$\underline{K}_3^{(i+1)} D_4^{(i)} + D_4^{(i)^T} \underline{K}_3^{(i+1)} + q_3^{(i)} = 0 \tag{3.74}$$

It can be seen that the complete reduced-order decomposition is achieved in (3.69)-(3.74), that is, one needs to solve four reduced-order Lyapunov and two reduced-order Sylvester equations.

The existence of the unique and bounded solutions of (3.63)-(3.68) is guaranteed by the stability of $D^{(i)}(\varepsilon)$, (Moerder and Calise, 1985a). Due to stability of $D_1^{(i)}(\varepsilon)$ and $D_4^{(i)}(\varepsilon)$ the unique solutions of (3.69)-(3.74) exist as well.

The zeroth-order solutions,

$$\underline{L}^{(i+1)} = \begin{bmatrix} \underline{L}_1^{(i+1)} & \varepsilon \underline{L}_2^{(i+1)} \\ \varepsilon \underline{L}_2^{(i+1)^T} & \underline{L}_3^{(i+1)} \end{bmatrix}$$

and

$$\underline{K}^{(i+1)} = \begin{bmatrix} \underline{K}_1^{(i+1)} & \varepsilon \underline{K}_2^{(i+1)} \\ \varepsilon \underline{K}_2^{(i+1)^T} & \underline{K}_3^{(i+1)} \end{bmatrix} \tag{3.75}$$

are $O(\varepsilon^2)$ close to the required ones $L^{(i+1)}$ and $K^{(i+1)}$ (see equations (3.63)-(3.68)). We can relate them through the error terms

$$L^{(i+1)} - \underline{L}^{(i+1)} = \varepsilon^2 M = \varepsilon^2 \begin{bmatrix} M_1 & \varepsilon M_2 \\ \varepsilon M_2^T & M_3 \end{bmatrix} \tag{3.76}$$

and

$$K^{(i+1)} - \underline{K}^{(i+1)} = \varepsilon^2 E = \varepsilon^2 \begin{bmatrix} E_1 & \varepsilon E_2 \\ \varepsilon E_2^T & E_3 \end{bmatrix} \tag{3.77}$$

Clearly the $O(\varepsilon^k)$ approximations for M's and E's will produce the $O(\varepsilon^{k+2})$ approximations of the required solutions. This is why we are interested in finding a convenient form for these error terms and the appropriate algorithm for their solution.

Using the results of Section 2.3.1 it can be shown that these error equations are given by

$$M_1 D_1^{(i)T} + D_1^{(i)}M_1 + D_2^{(i)}(\underline{L}_2^{(i+1)} + \varepsilon^2 M_2)^T +$$
$$+ (\underline{L}_2^{(i+1)} + \varepsilon^2 M_2)D_2^{(i)T} = 0 \tag{3.78}$$

$$M_2 D_4^{(i)T} + D_1^{(i)}M_2 + D_2^{(i)}M_3 + M_1 D_3^{(i)T} = 0 \tag{3.79}$$

$$M_3 D_4^{(i)T} + D_4^{(i)}M_3 + D_3^{(i)}(\underline{L}_2^{(i+1)} + \varepsilon^2 M_2)$$
$$+ (\underline{L}_2^{(i+1)} + \varepsilon^2 M_2)^T D_3^{(i)T} = 0 \tag{3.80}$$

and

$$E_1 D_1^{(i)} + D_1^{(i)T}E_1 + D_3^{(i)T}(\underline{K}_2^{(i+1)} + \varepsilon^2 E_2)^T$$
$$+ (\underline{K}_2^{(i+1)} + \varepsilon^2 E_2)D_3^{(i)} = 0 \tag{3.81}$$

$$E_2 D_4^{(i)} + D_1^{(i)T}E_2 + E_1 D_2^{(i)} + D_3^{(i)T}E_3 = 0 \tag{3.82}$$

$$E_3 D_4^{(i)} + D_4^{(i)T}E_3 + (\underline{K}_2^{(i+1)} + \varepsilon^2 E_2)^T D_2^{(i)} +$$
$$+ D_2^{(i)T}(\underline{K}_2^{(i+1)} + \varepsilon^2 E_2) = 0 \tag{3.83}$$

The weakly coupled and hierarchical structure of (3.78)-(3.83) can be exploited by proposing the recursive scheme, which leads, after some algebra, to the six low-order completely decoupled recursive equations

$$M_1^{(j+1)}D_1^{(i)T} + D_1^{(i)}M_1^{(j+1)} + D_2^{(i)}(L_2^{(i+1)} + \varepsilon^2 M_2^{(j)})^T +$$
$$+ (L_2^{(i+1)} + \varepsilon^2 M_2^{(j)})D_2^{(i)T} = 0 \tag{3.84}$$

$$M_2^{(j+1)}D_4^{(i)T} + D_1^{(i)}M_2^{(j+1)} + D_2^{(i)}M_3^{(j+1)} + M_1^{(j+1)}D_3^{(i)T} = 0 \tag{3.85}$$

$$M_3^{(j+1)}D_4^{(i)T} + D_4^{(i)}M_3^{(j+1)} + D_3^{(i)}(L_2^{(i+1)} + \varepsilon^2 M_2^{(j)})$$
$$+ (L_2^{(i+1)} + \varepsilon^2 M_2^{(j)})^T D_3^{(i)T} = 0 \tag{3.86}$$

and

$$E_1^{(j+1)}D_1^{(i)} + D_1^{(i)T}E_1^{(j+1)} + D_3^{(i)T}(K_2^{(i+1)} + \varepsilon^2 E_2^{(j)})^T$$
$$+ (K_2^{(i+1)} + \varepsilon^2 E_2^{(j)})D_3^{(i)} = 0 \tag{3.87}$$

$$E_2^{(j+1)}D_4^{(i)} + D_1^{(i)T}E_2^{(j+1)} + E_1^{(j+1)}D_2^{(i)} + D_3^{(i)T}E_3^{(j+1)} = 0 \tag{3.88}$$

$$E_3^{(j+1)}D_4^{(i)} + D_4^{(i)T}E_3^{(j+1)} + (K_2^{(i+1)} + \varepsilon^2 E_2^{(j)})^T D_2^{(i)} +$$
$$+ D_2^{(i)T}(K_2^{(i+1)} + \varepsilon^2 E_2^{(j)}) = 0 \tag{3.89}$$

with initial conditions being chosen as $M_1^{(0)} = E_1^{(0)} = 0$, $M_2^{(0)} = E_2^{(0)} = 0$ and $M_3^{(0)} = E_3^{(0)} = 0$.

Observe the decoupled structure of (3.84)-(3.89); the reduced-order Lyapunov equations (3.84), (3.86), (3.87) and (3.89) are solved first and then the Sylvester equations (3.85) and (3.88) are solved.

The existence of the unique solutions of (3.84)-(3.89) is guaranteed by the stability of $D_1^{(i)}(\varepsilon)$ and $D_4^{(i)}(\varepsilon)$. Note that iterations are performed with respect to j. For j = 0, (3.78) and (3.84) imply that

$$\left(M_1 - M_1^{(1)}\right)D_1^{(1)^T} + D_1^{(1)}\left(M_1 - M_1^{(1)}\right) = \varepsilon^2 \mathfrak{F}_1(M_2)$$

which by the stability of $D_1^{(1)}$ and the existence of the unique and bounded solution for M_2 gives

$$\| M_1 - M_1^{(1)} \| = O(\varepsilon^2) \tag{3.90}$$

By the same arguments, from (3.80) and (3.86) we have

$$\| M_3 - M_3^{(1)} \| = O(\varepsilon^2) \tag{3.91}$$

Subtracting (3.85) from (3.79) and using (3.90) and (3.91) lead to

$$\| M_2 - M_2^{(1)} \| = O(\varepsilon^2) \tag{3.92}$$

By analogy, (equations (3.81)-(3.83) and (3.87)-(3.89) have a similar form to (3.78)-(3.80) and (3.84)-(3.86)), equations (3.81)-(3.83) and (3.87)-(3.89) will produce

$$\| E_p - E_p^{(1)} \| = O(\varepsilon^2), \quad p = 1, 2, 3 \tag{3.93}$$

Continuing the same procedure for $j \geq 1$, it can be shown that

$$\| M_p - M_p^{(k)} \| = O(\varepsilon^{2k}), \quad p = 1, 2, 3 \tag{3.94}$$

and

$$\| E_p - E_p^{(k)} \| = O(\varepsilon^{2k}), \quad p = 1, 2, 3 \tag{3.95}$$

where k represents the number of iterations performed on the proposed algorithm.

Equations (3.94) and (3.95) imply

$$\| M - M^{(j)} \| = O(\varepsilon^{2j}), \qquad j = 1, 2, 3, \ldots \tag{3.96}$$

and

$$\| E - E^{(j)} \| = O(\varepsilon^{2j}), \qquad j = 1, 2, 3, \ldots \tag{3.97}$$

From (3.76)-(3.77) and (3.96)-(3.97) can be concluded that

$$\| L^{(i+1)} - \left(\underline{L}^{(i+1)} - M^{(j)} \right) \| = O(\varepsilon^{2j}), \quad j = 1, 2, 3, ... \tag{3.98}$$

and

$$\| K^{(i+1)} - \left(\underline{K}^{(i+1)} - E^{(j)} \right) \| = O(\varepsilon^{2j}), \quad j = 1, 2, 3, ...: \tag{3.99}$$

which in fact prove the following theorem.

Theorem 3.2 The algorithms (3.84)-(3.86) and (3.87)-(3.89) converge, for sufficiently small value of ε to the solution of the error terms, and thus to the required solutions $L^{(i+1)}$ and $K^{(i+1)}$, with the rate of convergence of $O(\varepsilon^2)$.

3.5 Case Study: Twelve Plate Absorption Column

In order to illustrate the efficiency of the proposed algorithm for weakly coupled systems, the method is applied to the mathematical model of the twelve-plate absorption column.

Today, chemical engineers are considerably interested in the use of modern control theory. In applying this theory, chemical processes are often approximated or described by linear state space models. In many cases, processes of interest to chemical engineers can be approximated by linear systems whose matrix A is tridiagonal. For example, many stagewise diffusion operations can be described by such models. From the analogy that exists among the processes of mass, heat and motion transfer, it can be concluded that the class of linear systems, in which the matrix A is tridiagonal, is quite common in practice. For that reason the control of those systems is of the particular interest.

The following is a brief description of the twelve plate absorption column, with changes permitted in the gas and liquid feed compositions. Here, the problem of the column control is obtaining a constant concentration of the outlet on the column, subject to "initial conditions" disturbances. The more detailed derivations and descriptions of the physical process can be found in (Lapidus and Amundson, 1950, Lapidus et. al. 1961, Petkovski, 1981). All numbers and parameters not specified can be found in the above references.

Using material balance relations around each plate and equilibrium relationships, the twelve-plate, absorption column may be described by a linear time-invariant state space model

$$A = \begin{bmatrix} A_1 & A_2 \\ A_3 & A_4 \end{bmatrix}$$

where

$$A_1 = A_4 = \begin{bmatrix} a_1 & a_2 & 0 & 0 & 0 & 0 \\ a_3 & a_1 & a_2 & 0 & 0 & 0 \\ 0 & a_3 & a_1 & a_2 & 0 & 0 \\ 0 & 0 & a_3 & a_1 & a_2 & 0 \\ 0 & 0 & 0 & a_3 & a_1 & a_2 \\ 0 & 0 & 0 & 0 & a_3 & a_1 \end{bmatrix}$$

$$\begin{bmatrix} B_1 & \varepsilon B_2 \end{bmatrix} = \begin{bmatrix} b_1 & 0 \\ 0 & 0 \\ 0 & 0 \\ 0 & 0 \\ 0 & 0 \\ 0 & 0 \end{bmatrix} \qquad \begin{bmatrix} \varepsilon B_3 & B_4 \end{bmatrix} = \begin{bmatrix} 0 & 0 \\ 0 & 0 \\ 0 & 0 \\ 0 & 0 \\ 0 & 0 \\ 0 & b_2 \end{bmatrix}$$

and

$$C = \begin{bmatrix} 1 & 0 & 0 & 0 & 0 & 0 & 0 & 0 & 0 & 0 & 0 & 0 \\ 0 & 1 & 0 & 0 & 0 & 0 & 0 & 0 & 0 & 0 & 0 & 0 \\ 0 & 0 & 0 & 0 & 0 & 0 & 0 & 0 & 0 & 0 & 1 & 0 \\ 0 & 0 & 0 & 0 & 0 & 0 & 0 & 0 & 0 & 0 & 0 & 1 \end{bmatrix}$$

A_2 has all entries equal to zero expect for $(A_2)_{6,1} = a_2$ and A_3 has all entries equal to zero except for $(A_3)_{1,6} = a_3$

The initial equilibrium state corresponds to pure liquid feed and a gas feed of 0.35 lb solute/lb inert, $u^T(t_0) = [0 \quad 0.35]$. The final desired equilibrium state corresponds to pure liquid feed and gas feed of 0.5 lb solute/lb inert, $u^T(t_f) = [0 \quad 0.5]$.

The initial conditions are

$$(x_1(0))^T = [-0.036 \quad -0.066 \quad -0.092 \quad -0.113 \quad -0.132 \quad -0.148]$$

$$(x_2(0))^T = [-0.161 \quad -0.173 \quad -0.182 \quad -0.190 \quad -0.197 \quad -0.203]$$

The entries in the matrices A and B are

$$a_1 = -1.73058, \qquad\qquad a_2 = 0.634231$$

$$a_3 = b_1 = 0.538827, \qquad\qquad b_2 = 0.8809$$

From the control theory point of view one of the basic features of an absorption column is a large number of stages. For that reason, optimization of a column using complete feedback, i.e. measuring all of plate concentrations, is not a practical solution. A realistic control scheme can be obtained by the application of output constrained regulators. In addition, the large dimensionality of these processes requires large amounts of computation for their solution. To overcome the computational difficulties in finding the output feedback control the weakly coupling approach could be used. The structure of the plant matrix suggests the application of this approach and the way of decoupling this matrix into submatrices. In this case a great simplification and reduction of the number of equations needed for calculation of output feedback control can be attained.

For matrices in the performance index, the matrix $Q = I_{12 \times 12}$ and $R = I_{2 \times 2}$. The small weak coupling parameter ε is chosen as $\varepsilon = 0.5$.

Figure 3.1 gives the schematic diagram of the considered column controlled via the output feedback regulator. A time response of a controlled and uncontrolled system is given in Figure 3.2 for the output variable of the plant $x_{12}(t)$. It is seen that the time response of the closed loop system is a significant improvement over the open loop system. On the other hand, the results presented in Figure 3.2 indicate that the control only using feedback from the measured variables will be sufficient in this case.

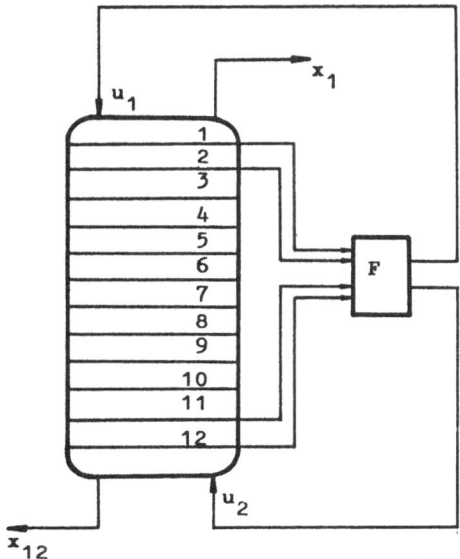

Fig 3.1 Twelve plate absorption column

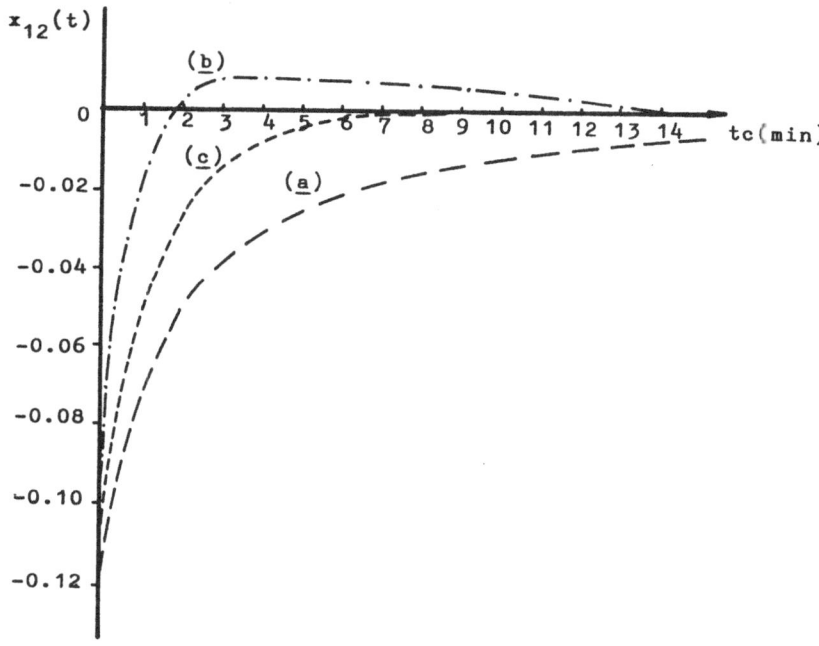

Fig. 3.2 The response of the variable $x_{12}(t)$

In Table 3.3, we present results for the criterion $J_{opt}^{(1)}$ and the gain error for the global algorithm (Moerder and Calise, 1985a), and the corresponding quantities for the proposed reduced-order recursive algorithm. The parameter α is chosen as $\alpha = 0.5$.

The initial value for the gain, $F^{(0)}$ is obtained by using the method proposed in (Petkovski and Rakić, 1978). The global algorithm takes 11 iterations to achieve the accuracy of up to 5 decimal digits, where $J_{opt} = 0.21273$.

It is important to note that the non-uniqueness of the solution of equations (3.7), (3.8) and (3.58) is shown by the entries in the fourth column of Table 3.3 which are obtained by using the global algorithm. It is seen that there are several possible solutions to the optimal control problem even though convergence to the optimal value of the criterion is achieved at $i = 11$.

Furthermore, for $i \geq 22$, with $\alpha = 0.5$, the global algorithm fails to produce the solution so that it can not converge to the unique value of the gain. From the entries in the fifth column of Table 3.3, it is clear that by using the reduced-order algorithm proposed, the difficulty of non-uniqueness of the solution to the optimal output control problem is resolved since the reduced-order algorithm produces a unique value of the feedback gain F. In addition, there are no problems with system instability when the reduced-order algorithm is used.

In order to facilitate finding the solution to the problem under study by using the global algorithm and to avoid problems with system instability, smaller values of α were used. The entries in Table 3.4 show the results obtained by using $\alpha = 0.05$ and $\alpha = 0.1$. The global algorithm fails to produce a unique value for the solution even though convergence to the optimal value of the criterion is achieved at $i = 116$ and $i = 55$ for $\alpha = 0.05$ and 0.1 respectively.

The example clearly shows the superiority of the reduced-order algorithm over the global algorithm.

$$\Delta_1 = \left\| F_{opt}^{(i)} - F_{opt} \right\|_\infty \ , \ \Delta_2 = \left\| F_{red}^{(i)} - F_{opt}^{(i)} \right\|_\infty$$

$\varepsilon = 0.5$ $\alpha = 0.5$ i	$J_{opt}^{(i)}$	$J_{red}^{(i)}$	Δ_1	Δ_2 $j = 6$
1	0.97305	0.97289	13.028	13.086
2	0.27731	0.27778	4.050	3.975
3	0.24112	0.24109	2.527	2.616
4	0.22316	0.22308	1.677	2.207
5	0.21596	0.21604	1.834	1.574
6	0.21355	0.21372	7.861	0.908
7	0.21286	0.21301	11.759	0.477
8	0.21277	0.21281	3.003	0.242
9	0.21274	0.21275	3.157	0.123
10	0.21274	0.21274	4.625	0.064
12	0.21273	0.21273	6.626	0.019
16	0.12173	0.21273	62.600	0.002
18	0.21273	0.21273	36.207	0.000
20	0.21273	0.21273	26.833	0.000
22	*	0.21273	*	0.000

* = global algorithm fails to produce solution for i = 22

Table 3.3 Optimal and approximate criteria and gains

	$\Delta_i = \left\| F_{opt}^{(i)} - F_{opt} \right\|_\infty$	
i	Δ_i	
	$\alpha = 0.05$	$\alpha = 0.1$
1	13.028	13.028
10	3.175	1.587
20	1.626	1.107
30	1.058	2.071
40	1.025	2.666
50	1.249	0.608
60	1.811	1.110
70	2.527	1.753
80	2.631	2.019
90	1.632	2.069
100	0.826	2.009
110	0.714	1.903
120	0.949	1.783
130	1.253	1.666
140	1.510	1.555
150	1.692	1.456
160	1.804	1.366
170	1.864	1.283
180	1.889	1.210
190	1.888	1.144
200	1.867	1.082

Table 3.4 Nonuniqueness of the global algorithm

LINEAR STOCHASTIC SYSTEMS

4.1 Recursive Approach to Singularly Perturbed Linear Stochastic Systems

Singularly perturbed linear stochastic estimation and control problems were studied in the past decade by a few researchers (Haddad, 1976; Haddad and Kokotović, 1977; Khalil and Gajić, 1984; Tenekatzis and Sandell, 1977). The recent paper (Khalil and Gajić, 1984) seems to be the most complete one. It alleviates the difficulties of the previous approaches and is conceptually simple. We shall briefly summarize the main results of (Khalil and Gajić, 1984). Consider the singularly perturbed system

$$\dot{x}_1 = A_1 x_1 + A_2 x_2 + B_1 u + G_1 w \tag{4.1}$$

$$\varepsilon \dot{x}_2 = A_3 x_1 + A_4 x_2 + B_2 u + G_2 w \tag{4.2}$$

$$y = C_1 x_1 + C_2 x_2 + v \tag{4.3}$$

where $x_1 \in R^{n_1}$ and $x_2 \in R^{n_2}$ are state vectors, $u \in R^m$ is a control input, $w \in R^{r_1}$ and $v \in R^{r_2}$ are zero-mean, stationary, white Gaussian noise with intensities $W > 0$ and $V > 0$ respectively, and ε is a small positive parameter. In the following A_i, B_j, G_j, C_j, $i = 1, \ldots, 4$, $j = 1, 2$, are constant matrices; in general, they are analytic functions of ε (Khalil and Gajić, 1984). With (4.1)-(4.3) consider the performance criterion

$$J = \lim_{\substack{t_0 \to -\infty \\ t_1 \to \infty}} \frac{1}{t_1 - t_0} E \left\{ \int_{t_0}^{t_1} \left[\begin{pmatrix} x_1 \\ x_2 \end{pmatrix}^T R_1 \begin{pmatrix} x_1 \\ x_2 \end{pmatrix} + u^T R_2 u \right] dt \right\} \tag{4.4}$$

with positive definite R_2 and positive semi-definite R_1, which has to be minimized.

The optimal control is given by

$$u = -F_1(\varepsilon)\hat{x}_1 - F_2(\varepsilon)\hat{x}_2 \tag{4.5}$$

where $\hat{x}_1$ and $\hat{x}_2$ are optimal estimates of the state vectors x_1 and x_2

$$\dot{\hat{x}}_1 = A_1\hat{x}_1 + A_2\hat{x}_2 + B_1u + K_1(\varepsilon)(y - C_1\hat{x}_1 - C_2\hat{x}_2) \tag{4.6a}$$

$$\varepsilon\dot{\hat{x}}_2 = A_3\hat{x}_1 + A_4\hat{x}_2 + B_2u + K_2(\varepsilon)(y - C_1\hat{x}_1 - C_2\hat{x}_2) \tag{4.6b}$$

The matrices F_1, F_2 and K_1, K_2 are regulator and filter gains respectively

$$F_1 = R_2^{-1}(B_1^TP_1 + B_2^TP_2^T), \quad F_2 = R_2^{-1}(\varepsilon B_1^TP_2 + B_2^TP_3) \tag{4.7a}$$

$$K_1 = (O_1C_1^T + O_2C_2^T)V^{-1}, \quad K_2 = (\varepsilon O_2^TC_1^T + O_3C_2^T)V^{-1} \tag{4.7b}$$

where P_i, Q_i, $i = 1, 2, 3$ are solutions of the corresponding regulator and filter Riccati equations

$$A^T(\varepsilon)P(\varepsilon) + P(\varepsilon)A(\varepsilon) - P(\varepsilon)S_R(\varepsilon)P(\varepsilon) + R_1 = 0 \tag{4.8a}$$

$$A(\varepsilon)Q(\varepsilon) + Q(\varepsilon)A^T(\varepsilon) - Q(\varepsilon){\cdot}S_F{\cdot}Q(\varepsilon) + G(\varepsilon)WG^T(\varepsilon) = 0 \tag{4.8b}$$

with scaling compatible to the nature of their solution

$$P(\varepsilon) = \begin{bmatrix} P_1 & \varepsilon P_2 \\ \varepsilon P_2^T & \varepsilon P_3 \end{bmatrix} \qquad Q(\varepsilon) = \begin{bmatrix} Q_1 & Q_2 \\ Q_2^T & \frac{1}{\varepsilon}Q_3 \end{bmatrix} \tag{4.9}$$

and newly defined matrices as

$$A(\varepsilon) = \begin{bmatrix} A_1 & A_2 \\ \dfrac{A_3}{\varepsilon} & \dfrac{A_4}{\varepsilon} \end{bmatrix}, \quad B(\varepsilon) = \begin{bmatrix} B_1 \\ \dfrac{B_2}{\varepsilon} \end{bmatrix}, \quad G(\varepsilon) = \begin{bmatrix} G_1 \\ \dfrac{G_2}{\varepsilon} \end{bmatrix} \qquad (4.10)$$

$$C = \begin{bmatrix} C_1, & C_2 \end{bmatrix}, \quad S_R(\varepsilon) = B(\varepsilon) R_2^{-1} B^T(\varepsilon), \quad S_F = C^T V^{-1} C$$

Eliminating u form (4.6), by using (4.5), the optimal filter can be represented as system driven by the innovation process $\nu = y - C_1 \hat{x}_1 - C_2 \hat{x}_2$

$$\dot{\hat{x}}_1 = (A_1 - B_1 F_1)\hat{x}_1 + (A_2 - B_1 F_2)\hat{x}_2 + K_1 \nu \qquad (4.11a)$$

$$\varepsilon \dot{\hat{x}}_2 = (A_3 - B_2 F_1)\hat{x}_1 + (A_4 - B_2 F_2)\hat{x}_2 + K_2 \nu \qquad (4.11b)$$

As was shown in (Khalil and Gajić, 1984), for the purpose of achieving decomposition on the slow and fast variables, this filter is transformed via the use of a nonsingular transformation (Chang, 1972) into new coordinates

$$\begin{bmatrix} \hat{\eta}_1 \\ \hat{\eta}_2 \end{bmatrix} = \begin{bmatrix} I - \varepsilon ML & -\varepsilon M \\ L & I \end{bmatrix} \begin{bmatrix} \hat{x}_1 \\ \hat{x}_2 \end{bmatrix} \qquad (4.12)$$

so that that the filter becomes

$$\dot{\hat{\eta}}_1 = \left[(A_1 - B_1 F_1) - (A_2 - B_1 F_2)L \right]\hat{\eta}_1 + (K_1 - MK_2 - \varepsilon MLK_1)\nu \qquad (4.13a)$$

$$\varepsilon \dot{\hat{\eta}}_2 = \left[(A_4 - B_2 F_2) + \varepsilon L(A_2 - B_1 F_2) \right]\hat{\eta}_2 + (K_2 + \varepsilon LK_1)\nu \qquad (4.13b)$$

with the innovation process $\nu = y - (C_1 - C_2 L)\hat{\eta}_1 - \left[C_2 + \varepsilon(C_1 - C_2 L)M \right]\hat{\eta}_2$.

The optimal control is now given by

$$u = -(F_1 - F_2 L)\hat{\eta}_1 - \left[F_2 + \varepsilon(F_1 - F_2 L)M \right]\hat{\eta}_2 \qquad (4.14)$$

Matrices L and M satisfy

$$\left(A_4 - B_2F_2\right)L - \left(A_3 - B_2F_1\right) - \varepsilon L\left[\left(A_1 - B_1F_1\right) - \left(A_2 - B_1F_2\right)L\right] = 0 \quad (4.15a)$$

$$-M\left(A_4 - B_2F_2\right) + \left(A_2 - B_1F_2\right) - \varepsilon ML\left(A_2 - B_1F_2\right)$$

$$+ \varepsilon\left[\left(A_1 - B_1F_1\right) - \left(A_2 - B_1F_2\right)L\right]M = 0 \quad (4.15b)$$

Thus, in order to find the optimal solution in the decomposed form above, we have to solve two Riccati equations (4.8), a weakly nonlinear equation (4.15a) and a linear equation (4.15b).

The following lemma is summarized from (Chow and Kokotović, 1976, Khalil and Gajić, 1984).

Lemma 4.1 If A_4 is nonsingular and the triples $\left(A_0, B_0, \rho_0\right)$, $\left(A_0, G_0, C_0\right)$, $\left(A_4, B_2, \rho_2\right)$, $\left(A_4, G_2, C_2\right)$ are stabilizable and detectable then, for a sufficiently small ε (4.8) have unique stabilizing solutions which possess power series expansions at $\varepsilon = 0$.

Matrices appearing in Lemma 4.1 are given by

$$A_0 = A_1 - A_2A_4^{-1}A_3 \qquad\qquad B_0 = B_1 - A_2A_4^{-1}B_2$$

$$G_0 = G_1 - A_2A_4^{-1}G_2 \qquad\qquad C_0 = C_1 - C_2A_4^{-1}A_3$$

$$R_1 = \left(\rho_1\ \rho_2\right)^T\left(\rho_1\ \rho_2\right) \qquad\qquad \rho_0 = \rho_1 - \rho_2A_4^{-1}A_3$$

Using results of Lemma 4.1 the approximate stabilizing control is defined as

$$u_{apr}^{(k)} = -\left(F_1^{(k)} - F_2^{(k)}L^{(k)}\right)\hat{\eta}_1^{(k)} - \left[F_2^{(k)} + \varepsilon\left(F_1^{(k)} - F_2^{(k)}L^{(k)}\right)M^{(k)}\right]\hat{\eta}_2^{(k)}$$

$$(4.16)$$

with approximative filters

$$\dot{\hat{\eta}}_1^{(k)} = \left[\left(A_1 - B_1F_1^{(k)}\right) - \left(A_2 - B_1F_2^{(k)}\right)L^{(k)}\right]\hat{\eta}_1^{(k)}$$

$$+ \left(K_1^{(k)} - M^{(k)}K_2^{(k)} - \varepsilon M^{(k)}L^{(k)}K_1^{(k)}\right)v^{(k)} \quad (4.17a)$$

$$\varepsilon\hat{\dot{\eta}}_2^{(k)} = \left[\left(A_4 - B_2 F_2^{(k)}\right) + \varepsilon L^{(k)}\left(A_2 - B_1 F_2^{(k)}\right)\right]\hat{\eta}_2^{(k)}$$

$$+ \left(K_2^{(k)} + \varepsilon L^{(k)} K_1^{(k)}\right)\nu^{(k)} \tag{4.17b}$$

where

$$\nu^{(k)} = y - \left(C_1 - C_2 L^{(k)}\right)\hat{\eta}_1^{(k)} - \left[C_2 + \varepsilon\left(C_1 - C_2 L^{(k)}\right)M^{(k)}\right]\hat{\eta}_2^{(k)}$$

and

$$F_1^{(k)} = R_2^{-1}\left(B_1^T P_1^{(k)} + B_2^T P_2^{T(k)}\right) = F_1 + O(\varepsilon^k)$$

$$F_2^{(k)} = R_2^{-1}\left(\varepsilon B_1^T P_2^{(k)} + B_2^T P_3^{(k)}\right) = F_2 + O(\varepsilon^k)$$

$$K_1^{(k)} = \left(Q_1^{(k)} C_1^T + Q_2^{(k)} C_2^T\right)V^{-1} = K_1 + O(\varepsilon^k)$$

$$K_2^{(k)} = \left(\varepsilon Q_2^{T(k)} C_1^T + Q_3^{(k)} C_2^T\right)V^{-1} = K_2 + O(\varepsilon^k)$$

The main result from (Khalil and Gajić, 1984) can be now summarized in the following theorem.

Theorem 4.1 Suppose that conditions of Lemma 4.1 hold. Let x_1 and x_2 be the optimal trajectories and J be the optimal value of the performance criterion. Let $x_1^{(k)}$, $x_2^{(k)}$, and $J^{(k)}$ be the corresponding quantities under the approximative control law $u_{apr}^{(k)}$ then

$$\frac{J^{(k)} - J}{J} = O(\varepsilon^k) \tag{4.18}$$

$$\mathrm{var}\left(x_1 - x_1^{(k)}\right) = O(\varepsilon^{2k}) \qquad \left(\text{as } t \to \infty\right) \tag{4.19a}$$

$$\mathrm{var}\left(x_2 - x_2^{(k)}\right) = O(\varepsilon^{2k-1}) \qquad \left(\text{as } t \to \infty\right) \tag{4.19b}$$

Note that by choosing appropriate initial conditions for (4.17) as $\hat{\eta}_1^{(k)}(0)$ $= \left(I - \varepsilon M^{(k)} L^{(k)}\right)\hat{x}_1(0) - \varepsilon M^{(k)}\hat{x}_2(0)$ and $\hat{\eta}_2^{(k)}(0) = L^{(k)}\hat{x}_1(0) + \hat{x}_2(0)$, then (4.19) holds for all $t \geq 0$.

Using standard techniques (Jamshidi, 1980), for the direct solution of Riccati equation (4.8) can be inappropriate since one would be faced with the stiff numerical problem of the full order. The well known singular perturbation techniques, (Kokotović and Yackel, 1972) based on the power series expansion with respect to ε, will convert given full order stiff problem (4.8), to the family of well defined reduced-order problems for which direct methods (Jamshidi, 1980) are very well suited. However, the power series expansion method is not recursive in its nature, and in the case when we are interested in a high degree of accuracy or when ε is not very small, which can often be the case, the size of the computations required can be considerable, even though we are solving low order problems. In such cases the advantage of doing series expansion method is questionable. The presence of a small parameter ε can be exploited from a different point of view, as was done in Chapter 2.

In summary, we can notice that the complete solution of LQG problem for singularly perturbed systems is based on the solution of the algebraic Riccati equations (4.8) and algebraic equations for the transformation matrices L and M, i.e., (4.15). Thus, the nature of their solutions determines the nature of the overall LQG problem. The algebraic Riccati equations can be solved using results from Section 2.2.2.

In order to illustrate numerical efficiency of the fixed point algorithm for solving the algebraic regulator Riccati equation consider the magnetic type control system example (Chow and Kokotović, 1976).

Example 4.1

$$
A(\varepsilon) = \begin{bmatrix} 0 & 0.4 & 0 & 0 \\ 0 & 0 & 0.345 & 0 \\ 0 & -\dfrac{0.524}{\varepsilon} & -\dfrac{-0.465}{\varepsilon} & \dfrac{0.262}{\varepsilon} \\ 0 & 0 & 0 & -\dfrac{1}{\varepsilon} \end{bmatrix}, \quad B(\varepsilon) = \begin{bmatrix} 0 \\ 0 \\ 0 \\ \dfrac{1}{\varepsilon} \end{bmatrix}
$$

$R_1 = \text{diag}\{1, 0, 1, 0\}$, $R_2 = 1$, $\varepsilon = 0.1$

With accuracy up to 6 decimal digits we got convergence to the exact solution in 4 iterations. Componentwise, results are given in Table 4.1

	$P = P^{(0)}$	$P^{(1)}$	$P^{(2)}$	$P^{(3)}$	$P^{(4)} = P^{(exact)}$
P_{11}	7.384024	7.540292	7.540064	7.540042	7.540043
P_{12}	5.904760	6.166314	6.170524	6.170445	6.170447
P_{13}	0.399308	0.405280	0.405345	0.405342	0.405342
P_{14}	0.100000	0.100000	0.100000	0.100000	0.100000
P_{22}	7.151604	7.452234	7.467309	7.467275	7.467278
P_{23}	0.379770	0.394804	0.395104	0.395100	0.395100
P_{24}	0.086123	0.089245	0.089202	0.089202	0.089202
P_{33}	0.104029	0.129797	0.130434	0.130441	0.130441
P_{34}	0.018036	0.024283	0.024398	0.024396	0.024396
P_{44}	0.004619	0.006183	0.006200	0.006200	0.006200

Table 4.1 Solution of Riccati equation for magnetic type control

Using different values of ε in the same example, we can note very good convergence rates even with relatively large values of perturbation parameter. These results are displayed in Table 4.2.

ε	No. of iterations for convergence
0.01	1
0.1	4
0.5	7
0.9	10

Table 4.2 Dependence of number of iterations on ε

The filter Riccati equation (4.8b) can be solved by the same algorithm taking into account the following analogies.

$$A_1 \rightarrow A_1^T, \ A_2 \rightarrow A_3^T, \ A_3 \rightarrow A_2^T, \ A_4 \rightarrow A_4^T, \ B_1 \rightarrow C_1^T, \ B_2 \rightarrow C_2^T,$$

$$\rho_1^T \rho_1 \rightarrow G_1 W G_1^T, \ \rho_1^T \rho_2 \rightarrow G_1 W G_2^T, \ \rho_2 \rho_2 \rightarrow G_2 W G_2^T, \ R_2 \rightarrow V$$

Thus, the remaining problem is to find a corresponding procedure for solving (4.15).

Solutions of L, M equations (4.15) that are needed for our LQG optimal control can be sought through the iterative form proposed in (Kokotović, Allemong, Winkelman and Chow, 1980).

$$L^{(i+1)} = \left(A_4 - B_2 F_2^{(N)}\right)^{-1}\left(A_3 - B_2 F_1^{(N)}\right) + \varepsilon\left(A_4 - B_2 F_2^{(N)}\right)^{-1}L^{(i)}\left[\left(A_1 - B_1 F_1^{(N)}\right)\right.$$

$$\left. - \left(A_2 - B_1 F_2^{(N)}\right)^{-1}L^{(i)}\right] \tag{4.20}$$

$$M^{(i+1)} = \left(A_2 - B_1 F_2^{(N)}\right)\left(A_4 - B_2 F_2^{(N)}\right)^{-1} - \varepsilon\left\{M^{(i)}L^{(N)}\left(A_2 - B_1 F_2^{(N)}\right)\right.$$

$$+ \left[\left(A_1 - B_1 F_1^{(N)}\right) - \left(A_2 - B_1 F_2^{(N)}\right)L^{(N)}\right]M^{(i)}\right\}\cdot\left(A_4 - B_2 F_2^{(N)}\right)^{-1}$$

$$i = 1, 2, ..., N\text{-}1 \tag{4.21}$$

with initial values

$$L^{(0)} = \left(A_4 - B_2 F_2^{(0)}\right)^{-1}\left(A_3 - B_2 F_1^{(0)}\right) \tag{4.22}$$

$$M^{(0)} = \left(A_2 - B_1 F_2^{(0)}\right)\left(A_4 - B_2 F_2^{(0)}\right)^{-1} \tag{4.23}$$

Note that in (4.20) $F_1^{(N)}$ and $F_2^{(N)}$ have already been determined using (4.7a) and the solution of the Riccati equation (4.8) which can be obtained using the algorithm from the previous section. The same holds for $L^{(N)}$ in equation (4.21).

Equations (4.20) and (4.21) have the form

$$D_3 L = \Upsilon_1 + \varepsilon \mathcal{F}_1(L, \varepsilon) \tag{4.24}$$

$$M D_3 = \Upsilon_2 + \varepsilon \mathcal{F}_2(M, L, \varepsilon) \tag{4.25}$$

where D_3 is a stable matrix, Υ_1 and Υ_2 are known constants and $\mathcal{F}_1$ and $\mathcal{F}_2$ are nonlinear functions. Then the following theorem holds for L and M.

Theorem 4.2. Under the conditions of Theorem 4.1, algorithm (4.20)-(4.21) converges to the exact solution L and M with the rate of convergence of $O(\varepsilon)$, i.e.

$$\| L - L^{(i+1)} \| = O(\varepsilon) \| L - L^{(i)} \|$$

$$\| M - M^{(i+1)} \| = O(\varepsilon) \| M - M^{(i)} \|$$

or equivalently

$$\| L - L^{(i)} \| = O(\varepsilon^i)$$

$$\| M - M^{(i)} \| = O(\varepsilon^i)$$

Proof of this theorem uses the same arguments as the proof of Theorem 2.2 and thus, it is omitted.

So far we have developed the iterative procedures that generate in a very efficient way all coefficients for the approximate solution of LQG (4.16)-(4.17). Those coefficients for an $O(\varepsilon^k)$ accuracy of the approximation of LQG are obtained by doing k-1 iterations on the same set of equations. Contrary to the power series expansions, where we have to solve k-different sets of equations, our iterative scheme is very useful in the case where a high order of accuracy is required. Instead of defining and solving a new set of equations, we have to perform just one additional iteration on the already existing set of equations. On the other hand in both cases we are faced with low-order numerical problems (in fact they are of the same order).

Equations (4.16)-(4.17) can be written in the following composite forms

$$u_{opt}^{(k)} = -f_1^{(k)} \hat{\eta}_1^{(k)} - f_2^{(k)} \hat{\eta}_2^{(k)} \tag{4.26}$$

$$\dot{\hat{\eta}}_1^{(k)} = a_1^{(k)} \hat{\eta}_1^{(k)} + g_1^{(k)} \nu^{(k)} \tag{4.27}$$

$$\varepsilon \dot{\hat{\eta}}_2^{(k)} = a_2^{(k)} \hat{\eta}_2^{(k)} + g_2^{(k)} \nu^{(k)} \tag{4.28}$$

where $\nu^{(k)} = y - \xi_1^{(k)} \hat{\eta}_1^{(k)} - \xi_2^{(k)} \hat{\eta}_2^{(k)}$, with obvious expressions for $f_i^{(k)}$, $a_i^{(k)}$, $g_i^{(k)}$, $\xi_i^{(k)}$, $i = 1, 2$. Then the suboptimal criterion

$$J^{(k)} = \lim_{\substack{t_0 \to -\infty \\ t_1 \to \infty}} \frac{1}{t_1 - t_0} E\left\{ \int_{t_0}^{t_1} \left[\left(\begin{matrix} x_1^{(k)} \\ x_2^{(k)} \end{matrix} \right)^T R_1 \left(\begin{matrix} x_1^{(k)} \\ x_2^{(k)} \end{matrix} \right) + u^{(k)T} R_2 u^{(k)} \right] dt \right\} \qquad (4.29)$$

is given by

$$J^{(k)} = tr\left\{ R_1 q_{11}^{(k)} + f^{(k)T} R_2 f^{(k)} q_{22}^{(k)} \right\}, \quad f^{(k)} = \left(f_1^{(k)T} f_2^{(k)T} \right)^T \qquad (4.30)$$

where

$$q_{11}^{(k)} = Var \left(\begin{matrix} x_1^{(k)} \\ x_2^{(k)} \end{matrix} \right) \qquad \text{and} \qquad q_{22}^{(k)} = Var \left(\begin{matrix} \hat{\eta}_1^{(k)} \\ \hat{\eta}_2^{(k)} \end{matrix} \right)$$

Quantities $q_{11}^{(k)}$ and $q_{22}^{(k)}$ can be obtained by studying the variance equation of the following system

$$\begin{bmatrix} \dot{x}_1^{(k)} \\ \varepsilon \dot{x}_2^{(k)} \\ \dot{\hat{\eta}}_1^{(k)} \\ \varepsilon \dot{\hat{\eta}}_2^{(k)} \end{bmatrix} = \begin{bmatrix} A_1 & A_2 & -B_1 f_1^{(k)} & -B_1 f_2^{(k)} \\ A_3 & A_4 & -B_2 f_1^{(k)} & -B_2 f_2^{(k)} \\ g_1^{(k)} C_1 & g_1^{(k)} C_2 & a_1^{(k)} - g_1^{(k)} \xi_1^{(k)} & -g_1^{(k)} \xi_2^{(k)} \\ g_2^{(k)} C_1 & g_2^{(k)} C_2 & -g_2^{(k)} \xi_1^{(k)} & a_2^{(k)} - g_2^{(k)} \xi_2^{(k)} \end{bmatrix} \begin{bmatrix} x_1^{(k)} \\ x_2^{(k)} \\ \hat{\eta}_1^{(k)} \\ \hat{\eta}_2^{(k)} \end{bmatrix}$$

$$+ \begin{bmatrix} G_1 & 0 \\ G_2 & 0 \\ 0 & g_1^{(k)} \\ 0 & g_2^{(k)} \end{bmatrix} \begin{bmatrix} w \\ v \end{bmatrix} \qquad (4.31)$$

or in a composite form

$$Z^{(k)} = \mathcal{A}^{(k)} Z^{(k)} + G^{(k)} \tilde{w} \qquad \qquad \tilde{w} = (w^T \ v^T)^T \qquad (4.32)$$

with obvious definitions for $\mathcal{A}^{(k)}$ and $G^{(k)}$.

The variance of $Z^{(k)}$ denoted by $q^{(k)}$ is given by the well-known Lyapunov equation

$$A^{(k)}q^{(k)} + q^{(k)}A^{(k)^T} + G^{(k)}\tilde{W}G^{(k)^T} = 0 , \quad \tilde{W} = \text{diag}(W, V) \qquad (4.33)$$

where $q^{(k)}$ is partitioned as

$$q^{(k)} = \begin{bmatrix} q_{11}^{(k)} & q_{12}^{(k)} \\ q_{12}^{(k)^T} & q_{22}^{(k)} \end{bmatrix} \qquad (4.34)$$

This procedure is demonstrated by a numerical example, showing the required convergence properties

$$f_i^{(k)} \to f_i^{opt}$$

$$a_i^{(k)} \to a_i^{opt}$$

$$g_i^{(k)} \to g_i^{opt}$$

$$\xi_i^{(k)} \to \xi_i^{opt}$$

$$J^{(k)} \to J^{opt}$$

$$k = 0, 1, ..., \text{ and } i = 1, 2.$$

4.2 CASE STUDY: F-8 Aircraft LQG Controller

In order to demonstrate the numerical behavior of the near-optimum design of singularly perturbed LQG regulators, we present results for an LQG controller of the F-8 aircraft which was considered in (Teneketzis and Sandell, 1977). The controller is designed to produce elevator commands to keep the aircraft in steady level flight in the face of wind disturbances. For simplicity, the wind disturbance is modeled as white. The aircraft's longitudinal variables are

$$x = \begin{bmatrix} V \\ \gamma \\ \alpha \\ q \end{bmatrix} \qquad\qquad u = \delta e$$

where

V	horizontal-velocity deviation (feet/second)
γ	flight-path angle (radians)
α	angle of attack (radians)
q	pitch rate (radians/second)
δ_e	the elevator deflection (radians)

The interpretation of these variables is given in Figure 4.1.

The equations of motion of the airplane are a set of coupled nonlinear equations in the longitudinal and lateral state variables. If the equations are linearized about the nominal state and control variables, then the resulting linear equations are found to approximately decouple into separate sets of the longitudinal and lateral dynamics. In our case the system model is given by

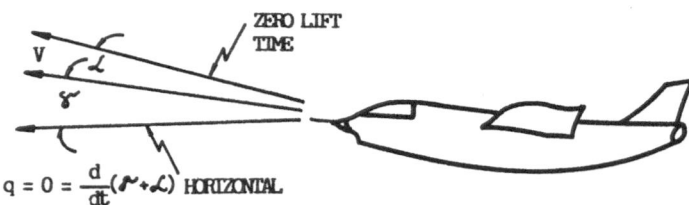

Figure 4.1 Aircraft longitudinal variables

$$
\begin{bmatrix} \dot{\tilde{x}}_1 \\ \dot{\tilde{x}}_2 \\ \dot{\tilde{x}}_3 \\ \dot{\tilde{x}}_4 \end{bmatrix} = \begin{bmatrix} -1.357 \times 10^{-2} & -32.2 & -46.3 & 0 \\ 1.2 \times 10^{-4} & 0 & 1.214 & 0 \\ -1.212 \times 10^{-4} & 0 & -1.214 & 1 \\ 5.7 \times 10^{-4} & 0 & -9.01 & -0.6696 \end{bmatrix} \begin{bmatrix} \tilde{x}_1 \\ \tilde{x}_2 \\ \tilde{x}_3 \\ \tilde{x}_4 \end{bmatrix} +
$$

$$
+ \begin{bmatrix} -0.433 \\ 0.1394 \\ -0.1394 \\ -0.1577 \end{bmatrix} u + \begin{bmatrix} -46.3 \\ 1.214 \\ -1.214 \\ -9.01 \end{bmatrix} w
$$

$$
y = \begin{bmatrix} 0 & 0 & 0 & 1 \\ 1 & 0 & 0 & 0 \end{bmatrix} \begin{bmatrix} \tilde{x}_1 \\ \tilde{x}_2 \\ \tilde{x}_3 \\ \tilde{x}_4 \end{bmatrix} + v
$$

where white noise processes w and v are independent and have intensities $W = 3.15 \times 10^{-4}$ and $V = \text{diag}[6.859 \times 10^{-4}, 40]$. The performance criterion is

$$
J = \lim_{\substack{t_0 \to -\infty \\ t_f \to \infty}} \frac{1}{t_f - t_0} E \left\{ \int_{t_0}^{t_f} [0.01\tilde{x}_1^2 + 3260(\tilde{x}_3^2 + \tilde{x}_4^2 + u^2)]dt \right\}
$$

The reader is referred to (Teneketzis and Sandell, 1977) for a discussion of the modeling aspects and the choice of J.

The open-loop eigenvalues are $-0.94 \pm j2.98$ and $-0.0075 \pm j0.076$ which shows clearly the two-time-scale property of the system. The choice of the state variables adopted in (Tenketzis and Sandell, 1977) led nicely to a formulation in which the first two variables are slow variables. A logical choice of the parameter ϵ is $\epsilon = 0.025$ which is roughly the ratio of the magnitude of the slow eigenvalues to the magnitude of the fast eigenvalues. The singularly perturbed nature of this system becomes more evident (Chow et.al. 1982) by using a state transformation $x = T \tilde{x}$ where

$$
T = \begin{bmatrix} 1 & 1618 & 133.92 & 200 \\ 0 & 500 & 40.8 & 61 \\ 0 & 0 & 600 & 0 \\ 0 & 0 & 0 & 200 \end{bmatrix}
$$

Introducing ε artificially by multiplying the left-hand sides by 0.025 the system takes the singularly perturbed form of (4.1)-(4.4) with

$$A_1 = \begin{bmatrix} 0.278386 & -0.965256 \\ 0.089833 & -0.290700 \end{bmatrix} \qquad A_2 = \begin{bmatrix} -0.074210 & 0.016017 \\ 0.012815 & -0.001398 \end{bmatrix}$$

$$A_3 = \begin{bmatrix} -0.001815 & 0.005873 \\ 0.002850 & -0.009223 \end{bmatrix} \qquad A_4 = \begin{bmatrix} -0.030344 & 0.075024 \\ -0.075092 & -0.016777 \end{bmatrix}$$

$$B_1 = \begin{bmatrix} 174.907714 \\ 54.392760 \end{bmatrix} \qquad B_2 = \begin{bmatrix} -2.091000 \\ -0.780500 \end{bmatrix}$$

$$P_1^T P_1 = \begin{bmatrix} 0.010000 & -0.032360 \\ -0.032360 & 0.104717 \end{bmatrix}$$

$$P_1^T P_2 = \begin{bmatrix} -0.000032 & -0.000130 \\ 0.000102 & 0.000421 \end{bmatrix}$$

$$P_2^T P_2 = \begin{bmatrix} 0.009056 & 0.000000 \\ 0.000000 & 0.081502 \end{bmatrix}$$

$$R_2 = 3260 \qquad W = 0.000315 \qquad V = \text{diag}\{0.000686, 40\}$$

$$C_1 = \begin{bmatrix} 0 & 0 \\ 1 & -3.236000 \end{bmatrix} \qquad C_2 = \begin{bmatrix} 0 & 0.005000 \\ -0.003152 & 0.013020 \end{bmatrix}$$

$$G_1 = \begin{bmatrix} -46.626960 \\ 7.858776 \end{bmatrix} \qquad G_2 = \begin{bmatrix} -18.210002 \\ -45.049998 \end{bmatrix}$$

Corresponding results are shown in Table 4.3

	FIRST	SECOND	THIRD	FOURTH	OPTIMAL
f_1	-0.000448	-0.000449	-0.000449	-0.000449	"
	0.002551	0.002554	0.002554	0.002554	
f_2	-0.000493	-0.000496	-0.000496	-0.000496	"
	-0.000061	-0.000065	-0.000065	-0.000065	
ξ_1	0.000230	0.000234	0.000235	0.000235	
	-0.000863	-0.000881	-0.000882	-0.000882	"
	0.999329	0.999314	0.999314	0.999314	
	-3.233571	-3.233517	-3.233516	-3.233516	
ξ_2	0.000004	0.000004	0.000004	0.000004	
	0.005009	0.005009	0.005009	0.005009	"
	0.010988	0.011113	0.011143	0.011143	
	0.019571	0.019609	0.019625	0.019626	
a_1	0.355782	0.355852	0.355843	0.355843	
	-1.409892	-1.410409	-1.410384	-1.410384	"
	0.114433	0.114482	0.114480	0.114480	
	0.429934	-0.430155	-0.430148	-0.430149	
a_2	-0.031306	-0.031313	-0.031314	-0.031314	
	0.074943	0.074936	0.074936	0.074036	"
	-0.075317	-0.075317	-0.075317	-0.075317	
	-0.016826	-0.016829	-0.016829	-0.016829	
g_1	25.459425	25.898515	25.897536	25.920266	
	0.000494	0.003132	0.004113	0.004134	"
	7.792644	7.945627	7.930356	7936231	
	0.001299	0.000704	0.000723	0.000723	
g_2	9.084229	9.104124	9.103953	9.103760	
	0.001998	0.001986	0.001972	0.001971	"
	22.483368	22.486045	22.486804.	22.486923	
	0.001873	0.001870	0.001875	0.001875	
J	25.066942	25.066604	25.066597	25.066597	25.066597

Table 4.3 A successive approximation solution of LQG for the aircraft F-8.

4.3 Recursive Approach to Weakly Coupled Linear Stochastic Systems

The linear quadratic Guassian control problem of weakly coupled systems has not been studied in the literature yet. Corresponding result for another class of small parameter systems - singularly perturbed systems (Kokotović and Khalil, 1986), is obtained in (Gajić, 1986) - by using the fixed point theory. Although the duality of the filter Riccati equation and regulator Riccati equation can be used together with results reported in (Gajić and Rayavarupu, 1989) to obtain corresponding approximations to the regulator and filter gains, such approximations will not be sufficient because they only reduce the off-line computations, but they do not help the on-line computations of implementing the Kalman filter which will be of the same order as the overall weakly coupled system. The weakly coupled structure of the global Kalman filter is exploited in this section such that it may be replaced by two lower order local filters. This has been achieved via the use of the decoupling transformation introduced in (Gajić and Shen, 1989a).

In this section we present the approach to the decomposition and approximation of the linear quadratic Gaussian (LQG) control problem of weakly coupled systems by treating the decomposition and approximation tasks separately from each other. The decoupling transformation of (Gajić and Shen, 1989a) is used for the exact block diagonalization of the global Kalman filter. The approximate feedback control law is then obtained by approximating the coefficients of the optimal local filters with the accuracy of $O(\varepsilon^N)$. The resulting feedback control law is shown to be a near optimal solution of the LQG by studying the corresponding closed loop system as a system driven by white noise. It is shown that the order of approximation of the optimal performance is $O(\varepsilon^N)$, and the order of approximation of the optimal system trajectories is $O(\varepsilon^{2N})$. All required coefficients of the desired accuracy are easily obtained by using the recursive fixed point type numerical techniques developed in Chapter 2. Given numerical algorithms converge to the required coefficients with the rate of convergence of $O(\varepsilon^2)$. In addition, only low-order subsystems are involved in the algebraic computations and no analyticity requirements are imposed on the system coefficients - which is the standard assumption in the power series expansion method. As a consequence of these, under very mild conditions (coefficients are bounded functions of the small coupling parameter) in addition to the standard stabilizability-detectability subsystem assumptions, we have achieved the reduction in both off-line and on-line computational requirements.

This section is organized as follows. At the beginning, we study the approximation of weakly coupled systems driven by white noise. It is shown that a Nth-order approximation in which the system coefficients are $O(\varepsilon^N)$ close to the exact ones is valid approximation in the sense that the differences between the exact and approximate solutions are $C(\varepsilon^{2N})$. Then, we use these results in the LQG study. A decoupling nonsingular transformation is used to represent the Kalman filter in new coordinates where local filters are completely decoupled. A Nth-order approximate feedback control law is defined by approximating coefficients by $O(\varepsilon^N)$. A study of the corresponding closed loop system, shows that the absolute increase in the performance criterion over its optimal value is $O(\varepsilon^N)$.

Consider the linear time-invariant weakly coupled system driven by white noise:

$$
\begin{bmatrix} \dot{x}_1 \\ \dot{x}_2 \end{bmatrix} = \begin{bmatrix} A_{11}(\varepsilon) & \varepsilon A_{12}(\varepsilon) \\ \varepsilon A_{21}(\varepsilon) & A_{22}(\varepsilon) \end{bmatrix} \begin{bmatrix} x_1 \\ x_2 \end{bmatrix} + \begin{bmatrix} G_{11}(\varepsilon) & \varepsilon G_{12}(\varepsilon) \\ \varepsilon G_{21}(\varepsilon) & G_{22}(\varepsilon) \end{bmatrix} \begin{bmatrix} w_1 \\ w_2 \end{bmatrix} \tag{4.35}
$$

where $x_i \in \mathbb{R}^{n_i}$, $w_i \in \mathbb{R}^{r_i}$, i = 1, 2, and ε is a small parameter. The system matrices are bounded functions of ε, (Gajić and Rayavarupu, 1989; Harkara, Petkovski and Gajić, 1989; Petrović and Gajić, 1988) of appropriate dimensions. The inputs $w_i(t)$ are zero mean, stationary. Gaussian uncorrelated white noise processes with intensities $W_i > 0$, i = 1, 2. It is well known that the variance of the linear systems driven by white noise is given by the Lyapunov equation (Kwakernaak and Sivan, 1972). In order to assure the existence of its solution we have assumed that $A_{ii}(\varepsilon)$, i = 1, 2 are stable matrices. The purpose of this section is to study approximations of $x_i(t)$, i = 1, 2 when ε is small. We are interested in approximations $x_i^N(t)$ which are defined by the following equations:

$$\begin{bmatrix} \dot{x}_1^N \\ \dot{x}_2^N \end{bmatrix} = \begin{bmatrix} A_{11}^N(\varepsilon) & \varepsilon A_{12}^N(\varepsilon) \\ \varepsilon A_{21}^N(\varepsilon) & A_{22}^N(\varepsilon) \end{bmatrix} \begin{bmatrix} x_1^N \\ x_2^N \end{bmatrix} + \begin{bmatrix} G_{11}^N(\varepsilon) & \varepsilon G_{12}^N(\varepsilon) \\ \varepsilon G_{21}^N(\varepsilon) & G_{22}^N(\varepsilon) \end{bmatrix} \begin{bmatrix} w_1 \\ w_2 \end{bmatrix} \qquad (4.36)$$

where

$$A_{ij}(\varepsilon) - A_{ij}{}^N(\varepsilon) = O(\varepsilon^N), \ i, \ j = 1, \ 2 \ .$$

$$G_{ij}(\varepsilon) - G_{ij}^N (\varepsilon) = O(\varepsilon^N), \ i, \ j = 1, \ 2 \qquad (4.37)$$

The quantities of interest are the variances of the errors

$$e_i(t) = x_i(t) - x_i^N(t), \ i = 1, \ 2 \qquad (4.38)$$

at steady state, and their impact on the quadratic form at steady state given by

$$\sigma = \text{tr} \left\{ \begin{bmatrix} H^T(\varepsilon)H(\varepsilon) & H^T(\varepsilon)J(\varepsilon) \\ J^T(\varepsilon)H(\varepsilon) & J^T(\varepsilon)J(\varepsilon) \end{bmatrix} E \begin{bmatrix} x_1(t)x_1^T(t) & x_1(t)x_2^T(t) \\ x_2(t)x_1^T(t) & x_2(t)x_2^T(t) \end{bmatrix} \right\} \qquad (4.39)$$

where $H(\varepsilon)$ and $J(\varepsilon)$ are bounded functions of ε also. Such quadratic forms will appear in the steady state LQG control problem. We examine the approximation of σ by σ^N given by

$$\sigma^N = \text{tr} \left\{ \begin{bmatrix} H^N{}^T(\varepsilon)H^N(\varepsilon) & H^N{}^T(\varepsilon)J^N(\varepsilon) \\ J^N{}^T(\varepsilon)H^N(\varepsilon) & J^N{}^T(\varepsilon)J^N(\varepsilon) \end{bmatrix} E \begin{bmatrix} x_1{}^N(t)x_1^{N^T}(t) & x_1{}^N(t)x_2^{N^T}(t) \\ x_2{}^N(t)x_1^{N^T}(t) & x_2{}^N(t)x_2^{N^T}(t) \end{bmatrix} \right\} \qquad (4.40)$$

where

$$H^N(\varepsilon) - H(\varepsilon) = O(\varepsilon^N), \ J^N(\varepsilon) - J(\varepsilon) = O(\varepsilon^N) \qquad (4.41)$$

In the following we will suppress the ε-dependence of the problem matrices in order to simplify notation.

The main results of this section are given in the following two theorems.

Theorem 4.3. Under stability assumptions of A_{ii}, $i = 1, 2$, the approximation errors at steady state satisfy

$$\text{Var}\{e_i\} = \text{Var}\{x_i - x_i^N\} = O(\varepsilon^{2N}), \; i = 1, 2$$

$$\text{Cov}\{e_1, e_2\} = O(\varepsilon^{2N})$$

(4.42)

Theorem 4.4. Under conditions stated in Theorem 4.3, the quadratic forms (4.39) and (4.40) at steady state satisfy

$$\Delta\sigma = \sigma - \sigma^N = O(\varepsilon^N)$$

(4.43)

The proof of these two theorems can be obtained by studying the following augmented system driven by white noise

$$\begin{bmatrix} \dot{x}_1 \\ \dot{e}_1 \\ \dot{x}_2 \\ \dot{e}_2 \end{bmatrix} = \begin{bmatrix} A_{11} & 0 & \varepsilon A_{11} & 0 \\ O(\varepsilon^N) & A_{11}{}^N & O(\varepsilon^{N+1}) & \varepsilon A_{12}{}^N \\ \varepsilon A_{21} & 0 & A_{22} & 0 \\ O(\varepsilon^{N+1}) & \varepsilon A_{21}{}^N & O(\varepsilon^N) & A_{22}{}^N \end{bmatrix} \begin{bmatrix} x_1 \\ e_1 \\ x_2 \\ e_2 \end{bmatrix} + \begin{bmatrix} G_{11} & \varepsilon G_{12} \\ O(\varepsilon^N) & O(\varepsilon^{N+1}) \\ \varepsilon G_{21} & G_{22} \\ O(\varepsilon^{N+1}) & O(\varepsilon^N) \end{bmatrix} \begin{bmatrix} w_1 \\ w_2 \end{bmatrix}$$

(4.44)

For shorthand notation (4.44) is rewritten as

$$\dot{z} = \Lambda z + Bw$$

(4.45)

with obvious definitions of z, w, Λ and B. The variance of z at steady state is given by the algebraic Lyapunov equation (Kwakernaak and Sivan, 1972)

$$0 = \Lambda Q + Q\Lambda^T + BWB^T$$

(4.46)

where

$$W = \begin{bmatrix} W_1 & 0 \\ 0 & W_2 \end{bmatrix}$$

and the variance of z is partitioned as

$$
Q = \begin{bmatrix}
Q_{11} & Q_{12} & \varepsilon Q_{13} & \varepsilon Q_{14} \\
Q_{12}{}^T & Q_{22} & \varepsilon Q_{23} & \varepsilon Q_{24} \\
\varepsilon Q_{13}{}^T & \varepsilon Q_{23}{}^T & Q_{33} & Q_{34} \\
\varepsilon Q_{14}{}^T & \varepsilon Q_{24}{}^T & Q_{34}{}^T & Q_{44}
\end{bmatrix}
$$

Studying (4.46) will produce after lengthy calculations

$$Q_{ij} = O(1), \qquad ij = 11, 13, 33 \tag{4.47a}$$

$$Q_{ij} = O(\varepsilon^{2N}) \quad ij = 22, 24, 44 \tag{4.47b}$$

$$Q_{ij} = O(\varepsilon^{N}), \quad ij = 12, 14, 23, 34 \tag{4.47c}$$

which proves Theorem 4.3.

Quadratic forms defined in (4.39) and (4.40) can be now expressed in terms of the elements of the matrix Q as

$$\sigma = \text{tr}\left\{ H^T H Q_{11} + J^T J Q_{33} + 2\varepsilon J^T H Q_{13} \right\} \tag{4.48}$$

and

$$
\begin{aligned}
\sigma^N = \text{tr}\Big\{ & H^{N^T} H^N (Q_{11} - 2Q_{12} + Q_{22}) + J^{N^T} J^N (Q_{33} - 2Q_{34} + Q_{44}) \\
& + 2\varepsilon J^{N^T} H^N (Q_{13} - Q_{23} - Q_{14} + Q_{24}) \Big\}
\end{aligned} \tag{4.49}
$$

From (4.48)-(4.49) and estimates for Q_{ij}, $i, j = 1, 2, 3, 4$, one has

$$\Delta\sigma = \sigma - \sigma^N = O(\varepsilon^N)\text{tr}\left\{ (Q_{11} + Q_{33}) + \left(H^{N^T} H^N + J^{N^T} J^N \right) \right\} + O(\varepsilon^{N+1}) \tag{4.50}$$

Since Q_{11}, Q_{33} and H^N, J^N are $O(1)$ quantities, one can conclude that $\Delta\sigma = O(\varepsilon^N)$, which completes the proof of Theorem 4.4

At this point we can introduce the linear quadratic Gaussian control problem of weakly coupled systems and study its approximation and decomposition by utilizing results from Theorems 4.3 and 4.4.

Consider the weakly coupled linear system

$$
\begin{bmatrix} \dot{x}_1 \\ \dot{x}_2 \end{bmatrix} = \begin{bmatrix} A_{11} & \varepsilon A_{12} \\ \varepsilon A_{21} & A_{22} \end{bmatrix} \begin{bmatrix} x_1 \\ x_2 \end{bmatrix} + \begin{bmatrix} B_{11} & \varepsilon B_{12} \\ \varepsilon B_{21} & B_{22} \end{bmatrix} \begin{bmatrix} u_1 \\ u_2 \end{bmatrix} + \begin{bmatrix} G_{11} & \varepsilon G_{12} \\ \varepsilon G_{21} & G_{22} \end{bmatrix} \begin{bmatrix} w_1 \\ w_2 \end{bmatrix} \tag{4.51}
$$

$$
\begin{bmatrix} y_1 \\ y_2 \end{bmatrix} = \begin{bmatrix} C_{11} & \varepsilon C_{12} \\ \varepsilon C_{21} & C_{22} \end{bmatrix} \begin{bmatrix} x_1 \\ x_2 \end{bmatrix} + \begin{bmatrix} v_1 \\ v_2 \end{bmatrix} \tag{4.52}
$$

where $x_i \in \mathbb{R}^{n_i}$, $u_i \in \mathbb{R}^{m_i}$, $y_i \in \mathbb{R}^{r_i}$, $i = 1, 2$ are state, control and measurement vectors respectively, and $w_i \in \mathbb{R}^{s_i}$, $v_i \in \mathbb{R}^{r_i}$, $i = 1, 2$ are independent zero-mean stationary white Gaussian noise processes with intensities W_i, V_i, $i = 1, 2$. The degree of interaction between subsystems is measured by a small parameter ε. With (4.51)-(4.52) consider the performance criterion

$$
J = \text{tr}\left\{ D^T D \ E \begin{bmatrix} x_1 x_1^T & x_1 x_2^T \\ x_2 x_1^T & x_2 x_2^T \end{bmatrix} + R \ E \begin{bmatrix} u_1 u_1^T & u_1 u_2^T \\ u_2 u_1^T & u_2 u_2^T \end{bmatrix} \right\} \tag{4.53}
$$

with positive definite R which has to be minimized at steady state. In the following all matrices are bounded functions of ε, (Gajić and Rayavarupu, 1989, Harkara, Petkovski and Gajić, 1989, Petrović and Gajić, 1988), of the appropriate dimensions. In addition, matrices $D^T D$ and R have the weakly coupled structure, namely, we have assumed that they are given by

$$
D^T D = \begin{bmatrix} D_1^T D_1 & \varepsilon D_1^T D_2 \\ \varepsilon D_2^T D_1 & D_2^T D_2 \end{bmatrix} \quad , \quad R = \begin{bmatrix} R_1 & 0 \\ 0 & R_2 \end{bmatrix}
$$

where $R_i \in \mathbb{R}^{m_i \times m_i}$ and $D_i^T D_i \in \mathbb{R}^{n_i \times n_i}$, $i = 1, 2$.

The optimal control law has the very well-known form (Kwakernaak and Sivan, 1972)

$$
\begin{bmatrix} u_1(t) \\ u_2(t) \end{bmatrix} = - \begin{bmatrix} F_{11} & \varepsilon F_{12} \\ \varepsilon F_{21} & F_{22} \end{bmatrix} \begin{bmatrix} \hat{x}_1(t) \\ \hat{x}_2(t) \end{bmatrix} \tag{4.54}
$$

$$\begin{bmatrix} \dot{\hat{x}}_1(t) \\ \dot{\hat{x}}_2(t) \end{bmatrix} = \begin{bmatrix} A_{11} & \varepsilon A_{12} \\ \varepsilon A_{21} & A_{22} \end{bmatrix} \begin{bmatrix} \hat{x}_1(t) \\ \hat{x}_2(t) \end{bmatrix} + \begin{bmatrix} B_{11} & \varepsilon B_{12} \\ \varepsilon B_{21} & B_{22} \end{bmatrix} \begin{bmatrix} u_1 \\ u_2 \end{bmatrix}$$

$$+ \begin{bmatrix} K_{11} & \varepsilon K_{12} \\ \varepsilon K_{21} & K_{22} \end{bmatrix} \begin{bmatrix} y_1 - C_{11}\hat{x}_1 - \varepsilon C_{12}\hat{x}_2 \\ y_2 - \varepsilon C_{21}\hat{x}_1 - C_{22}\hat{x}_2 \end{bmatrix} \qquad (4.55)$$

Introducing a notation

$$A = \begin{bmatrix} A_{11} & \varepsilon A_{12} \\ \varepsilon A_{21} & A_{22} \end{bmatrix}, \quad B = \begin{bmatrix} B_{11} & \varepsilon B_{12} \\ \varepsilon B_{21} & B_{22} \end{bmatrix}, \quad G = \begin{bmatrix} G_{11} & \varepsilon G_{12} \\ \varepsilon G_{21} & G_{22} \end{bmatrix}$$

$$C = \begin{bmatrix} C_{11} & \varepsilon C_{12} \\ \varepsilon C_{21} & C_{22} \end{bmatrix}, \quad F = \begin{bmatrix} F_{11} & \varepsilon F_{12} \\ \varepsilon F_{21} & F_{22} \end{bmatrix}, \quad K = \begin{bmatrix} K_{11} & \varepsilon K_{12} \\ \varepsilon K_{21} & K_{22} \end{bmatrix}$$

$$W = \begin{bmatrix} W_1 & 0 \\ 0 & W_2 \end{bmatrix}, \quad V = \begin{bmatrix} V_1 & 0 \\ 0 & V_2 \end{bmatrix}$$

the regulator and filter gains are obtained from

$$F = R^{-1}B^T P, \quad K = QC^T V^{-1} \qquad (4.56)$$

where P and Q are positive semi-definite stabilizing solutions of the algebraic Riccati equations

$$A^T P + PA - PS_R P + D^T D = 0, \quad S_R = BR^{-1}B^T \qquad (4.57)$$

$$AQ + QA^T - QS_F Q + GWG^T = 0, \quad S_F = C^T V^{-1}C \qquad (4.58)$$

Due to weakly coupled structure of all coefficients in (4.57)-(4.58), solutions of these equations have the form

$$P = \begin{bmatrix} P_1 & \varepsilon P_2 \\ \varepsilon P_2^T & P_3 \end{bmatrix} \qquad Q = \begin{bmatrix} Q_1 & \varepsilon Q_2 \\ \varepsilon Q_2^T & Q_3 \end{bmatrix} \qquad (4.59)$$

Solutions of (4.57)-(4.58) can be found in terms of the reduced-order problems by imposing standard stabilizability-detectability assumptions on subsystems. The efficient fixed point algorithm for solving (4.57) and (4.58) is obtained in Section 2.3.2. The algorithms for solving regulator and filter algebraic Riccati equations of weakly coupled systems are convergent under the following assumptions.

Assumption 4.1. Triples (A_{ii}, B_{ii}, D_{ii}), $i = 1, 2$ are stabilizable and detectable.

Assumption 4.2. Triples (A_{ii}, C_{ii}, G_{ii}), $i = 1, 2$ are stabilizable and detectable.

Getting approximate solutions for P and Q in terms of the reduced-order problems will produce saving in off-line computations. However, in the case of stochastic systems, where an additional dynamic system - filter - has to be built, one is particularly interested in the reduction of on-line computations. We will achieve that by the use of the decoupling transformation introduced in Section 2.4.

The Kalman filter (4.55) is viewed as a system driven by the innovation process. However, one might study the filter form when it is driven by both measurements and controls. The filter form under consideration is obtained from (4.55) as

$$\begin{bmatrix} \dot{\hat{x}}_1 \\ \dot{\hat{x}}_2 \end{bmatrix} = \begin{bmatrix} \left(A_{11} - B_{11}F_{11} - \varepsilon^2 B_{12}F_{12}\right) & \varepsilon\left(A_{12} - B_{11}F_{12} - B_{12}F_{22}\right) \\ \varepsilon\left(A_{21} - B_{21}F_{11} - B_{22}F_{21}\right) & \left(A_{22} - B_{22}F_{22} - \varepsilon^2 B_{21}F_{12}\right) \end{bmatrix} \begin{bmatrix} \hat{x}_1 \\ \hat{x}_2 \end{bmatrix}$$

$$+ \begin{bmatrix} K_{11} & \varepsilon K_{12} \\ \varepsilon K_{21} & K_{22} \end{bmatrix} \begin{bmatrix} \nu_1 \\ \nu_2 \end{bmatrix} \tag{4.60}$$

with innovation processes

$$\nu_1 = y_1 - C_{11}\hat{x}_1 - \varepsilon C_{12}\hat{x}_2$$

$$\nu_2 = y_2 - \varepsilon C_{21}\hat{x}_1 - C_{22}\hat{x}_2 \tag{4.61}$$

The nonsingular state transformation will block diagonalize (4.60) under condition that $(A_{11}-B_{11}F_{11}-\epsilon^2 B_{12}F_{21})$ and $(A_{22}-B_{22}F_{22}-\epsilon^2 B_{21}F_{12})$ have no eigenvalues in common (see Section 2.4). This transformation is given by

$$
\begin{bmatrix} \hat{\eta}_1 \\ \hat{\eta}_2 \end{bmatrix} = \begin{bmatrix} I - \epsilon^2 LH & -\epsilon L \\ \epsilon H & I \end{bmatrix} \begin{bmatrix} \hat{x}_1 \\ \hat{x}_2 \end{bmatrix} = T^{-1} \begin{bmatrix} \hat{x}_1 \\ \hat{x}_2 \end{bmatrix}
\tag{4.62}
$$

with

$$
T = \begin{bmatrix} I & \epsilon L \\ -\epsilon H & I - \epsilon^2 LH \end{bmatrix}
\tag{4.63}
$$

where matrices L and H satisfy equations given in Section 2.4. The optimal feedback control expressed in the new coordinates has the from

$$
u_1 = -f_{11}\hat{\eta}_1 - \epsilon f_{12}\hat{\eta}_2
\tag{4.64a}
$$

$$
u_2 = - \epsilon f_{21}\hat{\eta}_1 - f_{22}\hat{\eta}_2
\tag{4.64b}
$$

with

$$
\dot{\hat{\eta}}_1 = \alpha_1 \hat{\eta}_1 + \beta_{11}\nu_1 + \epsilon\beta_{12}\nu_2
\tag{4.65a}
$$

$$
\dot{\hat{\eta}}_2 = \alpha_2 \hat{\eta}_2 + \epsilon\beta_{21}\nu_1 + \beta_{22}\nu_2
\tag{4.65b}
$$

where

$$
f_{11} = F_{11} - \epsilon^2 F_{12}H, \qquad\qquad f_{12} = F_{12} + (F_{11} - \epsilon^2 F_{12}H)L
$$

$$
f_{21} = F_{21} - F_{22}H, \qquad\qquad f_{22} = F_{22} + \epsilon^2(F_{21} - F_{22}H)L
$$

$$
\alpha_1 = a_{11} - \epsilon^2 a_{12}H, \qquad\qquad \alpha_2 = a_{22} + \epsilon^2 Ha_{12}
$$

$$
\beta_{11} = K_{11} - \epsilon^2(LH + LK_{21}), \qquad\qquad \beta_{12} = K_{12} - LK_{22} - \epsilon^2 LHK_{12}
\tag{4.66}
$$

$$
\beta_{21} = HK_{11} + K_{21}, \qquad\qquad \beta_{22} = K_{22} + \epsilon^2 HK_{12}
$$

and

$$a_{11} = \left(A_{11} - B_{11}F_{11} - \varepsilon^2 B_{12}F_{21} \right)$$

$$a_{12} = \left(A_{12} - B_{11}F_{12} - B_{12}F_{22} \right)$$

$$a_{21} = \left(A_{21} - B_{21}F_{11} - B_{22}F_{21} \right)$$

$$a_{22} = \left(A_{22} - B_{22}F_{22} - \varepsilon^2 B_{21}F_{12} \right)$$

The innovation processes v_1 and v_2 are now given by

$$v_1 = y_1 - d_{11}\hat{\eta}_1 - \varepsilon d_{12}\hat{\eta}_2 \tag{4.67a}$$

$$v_2 = y_2 - \varepsilon d_{21}\hat{\eta}_1 - d_{22}\hat{\eta}_2 \tag{4.67b}$$

where

$$d_{11} = C_{11} - \varepsilon^2 C_{12}H, \qquad\qquad d_{12} = C_{11}L + C_{12} - \varepsilon^2 C_{12}HL,$$

$$d_{21} = C_{21} - C_{22}H, \qquad\qquad d_{22} = C_{22} + \varepsilon^2 (C_{21} - C_{22}H)L$$

Approximate control laws are defined by perturbing coefficients F_{ij}, K_{ij}, (i, j = 1, 2), L and H by $O(\varepsilon^k)$, k = 1, 2,.... , in other words by using k-th approximations for these coefficients, where k stands for the required order of accuracy, that is

$$u_1^{(k)} = - f_{11}^{(k)}\hat{\eta}_1^{(k)} - \varepsilon f_{12}^{(k)}\hat{\eta}_2^{(k)} \tag{4.68a}$$

$$u_2^{(k)} = - \varepsilon f_{21}^{(k)}\hat{\eta}_1^{(k)} - f_{22}^{(k)}\hat{\eta}_2^{(k)} \tag{4.68b}$$

with

$$\dot{\hat{\eta}}_1^{(k)} = \alpha_1^{(k)}\hat{\eta}_1^{(k)} + \beta_{11}^{(k)}v_1^{(k)} + \varepsilon \beta_{12}^{(k)}v_2^{(k)} \tag{4.69a}$$

$$\dot{\hat{\eta}}_2^{(k)} = \alpha_2^{(k)}\hat{\eta}_2^{(k)} + \varepsilon \beta_{21}^{(k)}v_1^{(k)} + \beta_{22}^{(k)}v_2^{(k)} \tag{4.69b}$$

where

$$v_1^{(k)} = y_1^{(k)} - d_{11}^{(k)}\hat{\eta}_1^{(k)} - \varepsilon d_{12}^{(k)}\hat{\eta}_2^{(k)} \tag{4.70a}$$

$$v_2^{(k)} = y_2^{(k)} - \varepsilon d_{21}^{(k)}\hat{\eta}_1^{(k)} - d_{22}^{(k)}\hat{\eta}_2^{(k)} \tag{4.70b}$$

and

$$f_{ij}^{(k)} = f_{ij} + O(\varepsilon^k), \qquad\qquad d_{ij}^{(k)} = d_{ij} + O(\varepsilon^k)$$

$$\beta_{ij}^{(k)} = \beta_{ij} + O(\varepsilon^k) \tag{4.71}$$

$$\alpha_{ij}^{(k)} = \alpha_{ij} + O(\varepsilon^k) \qquad\qquad i, j = 1, 2$$

The near-optimality of the proposed control law (4.68) is established in the following theorem.

Theorem 4.5. Let x_1 and x_2 be the optimal trajectories and J be the optimal value of the performance criterion. Let $x_1^{(k)}$, $x_2^{(k)}$ and $J^{(k)}$ be the corresponding quantities under the approximate control law $u^{(k)}$, then

$$J - J^{(k)} = O(\varepsilon^k) \qquad\qquad k = 0, 1, 2, \dots \tag{4.72}$$

$$\mathrm{Var}\{(x_i - x_i^{(k)})\} = O(\varepsilon^{2k}) \qquad\qquad k = 0, 1, 2, \dots \tag{4.73}$$

Proof of Theorem 4.5: The results of Theorems 4.3 and 4.4 are employed by studying system of equations driven by white noise. For the optimal control consider the equations

$$
\begin{bmatrix} \dot{\hat{\eta}}_1 \\ \dot{e}_1 \\ \dot{\hat{\eta}}_2 \\ \dot{e}_2 \end{bmatrix} =
\begin{bmatrix} \Lambda_{11} & \varepsilon\Lambda_{12} \\ \varepsilon\Lambda_{21} & \Lambda_{22} \end{bmatrix}
\begin{bmatrix} \hat{\eta}_1 \\ e_1 \\ \hat{\eta}_2 \\ e_2 \end{bmatrix} +
\begin{bmatrix} \Theta_{11} & \varepsilon\Theta_{12} \\ \varepsilon\Theta_{21} & \Theta_{22} \end{bmatrix}
\begin{bmatrix} v_1 \\ w_1 \\ v_2 \\ w_2 \end{bmatrix} \tag{4.74}
$$

where $e_i = \eta_i - \hat{\eta}_i$, $i = 1, 2$ are estimation errors. The corresponding equation for the approximate control is

$$
\begin{bmatrix} \dot{\hat{\eta}}_1^N \\ \dot{e}_1^N \\ \dot{\hat{\eta}}_2^N \\ \dot{e}_2^N \end{bmatrix} =
\begin{bmatrix} \Lambda_{11}^N & \varepsilon\Lambda_{12}^N \\ \varepsilon\Lambda_{21}^N & \Lambda_{22}^N \end{bmatrix}
\begin{bmatrix} \hat{\eta}_1^N \\ e_1^N \\ \hat{\eta}_2^N \\ e_2^N \end{bmatrix} +
\begin{bmatrix} \Theta_{11}^N & \varepsilon\Theta_{12}^N \\ \varepsilon\Theta_{21}^N & \Theta_{22}^N \end{bmatrix}
\begin{bmatrix} v_1 \\ w_1 \\ v_2 \\ w_2 \end{bmatrix} \tag{4.75}
$$

where $e_i^N = \eta_i - \hat{\eta}_i^N$ are corresponding estimation errors. The matrices Λ_{ij}, Θ_{ij} and Λ_{ij}^N, Θ_{ij}^N in (4.74) and (4.75) are obtained in obvious way. It can be verified that

$$\Lambda_{ij} - \Lambda_{ij}^N = O(\varepsilon^N), \qquad \Theta_{ij} - \Theta_{ij}^N = O(\varepsilon^N), \qquad i, j = 1, 2$$

and that $\Lambda_{ii}(0)$, $i = 1, 2$ are given by

$$\Lambda_{ii}(0) = \begin{bmatrix} A_{ii} - B_{ii}F_{ii} & K_{ii}C_{ii} \\ 0 & A_{ii} - K_{ii}C_{ii} \end{bmatrix} \tag{4.76}$$

which by stabilizability-detectability assumptions imposed on triples (A_{ii}, B_{ii}, D_i) and (A_{ii}, C_{ii}, G_{ii}), $i = 1, 2$, guarantees the stability of matrices $\Lambda_{ii}(0)$. The results of Theorems 4.3 and 4.4 can be now directly used to establish (4.72) and (4.73).

4.4 Case Study: Electric Power System Example

In order to demonstrate the numerical behavior of the near-optimum design of weakly coupled LQG regulator, we present results for an LQG controller of a power system composed of two interconnected areas (Geromel and Peres, 1985). The system model is given by (see Appendix 4.1)

$$A = \begin{bmatrix}
0.0 & 0.55 & 0.0 & 0.0 & 0.0 & -0.55 & 0.0 & 0.0 & 0.0 \\
0.0 & 0.0 & 1.0 & 0.0 & 0.0 & 0.0 & 0.0 & 0.0 & 0.0 \\
0.0 & -3.3 & -0.05 & 6.0 & 0.0 & 3.3 & 0.0 & 0.0 & 0.0 \\
0.0 & 0.0 & 0.0 & -3.3 & 3.3 & 0.0 & 0.0 & 0.0 & 0.0 \\
0.0 & 0.0 & -5.2 & 0.0 & -13.0 & 0.0 & 0.0 & 0.0 & 0.0 \\
0.0 & 0.0 & 0.0 & 0.0 & 0.0 & 0.0 & 1.0 & 0.0 & 0.0 \\
0.0 & 3.3 & 0.0 & 0.0 & 0.0 & -3.3 & -0.05 & 6.0 & 0.0 \\
0.0 & 0.0 & 0.0 & 0.0 & 0.0 & 0.0 & 0.0 & -3.3 & 3.3 \\
0.0 & 0.0 & 0.0 & 0.0 & 0.0 & 0.0 & -5.2 & 0.0 & -13.0
\end{bmatrix}$$

$$B^T = \begin{bmatrix} 0.0 & 0.0 & 0.0 & 0.0 & 13.0 & 0.0 & 0.0 & 0.0 & 0.0 \\ 0.0 & 0.0 & 0.0 & 0.0 & 0.0 & 0.0 & 0.0 & 0.0 & 13.0 \end{bmatrix}$$

$$C = \begin{bmatrix} 1.0 & 0.43 & 0.0 & 0.0 & 0.0 & 0.0 & 0.0 & 0.0 & 0.0 \\ 0.0 & 0.0 & 0.0 & 1.0 & 0.0 & 0.0 & 0.0 & 0.0 & 0.0 \\ -1.0 & 0.0 & 0.0 & 0.0 & 0.0 & 0.43 & 0.0 & 0.0 & 0.0 \\ 0.0 & 0.0 & 0.0 & 0.0 & 0.0 & 0.0 & 0.0 & 1.0 & 0.0 \end{bmatrix}$$

$$D^T D = \begin{bmatrix} 1.0 & 0.0 & 0.0 & 0.0 & 0.0 & 0.0 & 0.0 & 0.0 & 0.0 \\ 0.0 & 1.3 & 0.0 & 0.0 & 0.0 & -0.3 & 0.0 & 0.0 & 0.0 \\ 0.0 & 0.0 & 1.0 & 0.0 & 0.0 & 0.0 & 0.0 & 0.0 & 0.0 \\ 0.0 & 0.0 & 0.0 & 0.0 & 0.0 & 0.0 & 0.0 & 0.0 & 0.0 \\ 0.0 & 0.0 & 0.0 & 0.0 & 0.0 & 0.0 & 0.0 & 0.0 & 0.0 \\ 0.0 & -0.3 & 0.0 & 0.0 & 0.0 & 1.3 & 0.0 & 0.0 & 0.0 \\ 0.0 & 0.0 & 0.0 & 0.0 & 0.0 & 0.0 & 1.0 & 0.0 & 0.0 \\ 0.0 & 0.0 & 0.0 & 0.0 & 0.0 & 0.0 & 0.0 & 0.0 & 0.0 \\ 0.0 & 0.0 & 0.0 & 0.0 & 0.0 & 0.0 & 0.0 & 0.0 & 0.0 \end{bmatrix}$$

$$R = \begin{bmatrix} 1.0 & 0 \\ 0 & 1.0 \end{bmatrix}$$

It is assumed that $G = B$, and that the noise intensity matrices are given by

$$W_1 = 0.1, \ W_2 = 0.1, \ V_1 = I_2, \ V_2 = I_2$$

We can note relatively big elements in the cross coupling matrices A_{12}, A_{21} and especially C_{21}. The small parameter ε is built in the problem. The value for ε should be estimated from the problem strongest coupled matrix - in this case matrix C. It seems from our experience that the formula

$$\varepsilon = \frac{\max(\ \|\ C_{12}\ \|. \ \|\ C_{21}\ \|\ )}{\min(\ \|\ C_{11}\ \|. \ \|\ C_{22}\ \|\ )} = \frac{1}{1.43} = 0.699 \tag{4.77}$$

produces quite good estimate for ε, where $\|\ \ \|$ is any suitable norm.

The simulation results are presented in the following table

k	$J^{(k)}$	$J^{(k)} - J$	$(0.7)^k$
2	∞	∞	*
4	∞	∞	*
6	5.9415	0.9645	0.11765
10	5.1111	0.1341	0.02825
18	4.9788	0.0018	0.00163
26	4.9770	$< 10^{-4}$	9.4×10^{-5}
Optimal	4.9770	*	*

Table 4.4. Approximate values for criterion

The small parameter ε is relatively big in this example, that is $\varepsilon = 0.7$. Since $O(0.7^{26}) \approx 10^{-4}$, it will require 24 terms in order to get the accuracy of 10^{-4} if the power series expansion method is used - which is not feasible. On the other hand, the fixed point method scheme used in this section will demand 12 iterations (rate of convergence is $O(\varepsilon^2)$) of the presented algorithms - which can be easily achieved. Even more, it happens in this problem that the $O(\varepsilon^2)$ and $O(\varepsilon^4)$ approximate filters do not stabilize the plant-filter augmented system, and the approximate filter has to be found with the accuracy of at least $O(\varepsilon^6)$.

Table 4.4 verifies the result of Theorem 4.5, namely $J - J^{(k)} = O(\varepsilon^k)$, and support the formula (4.77) for the estimate of the weakly coupling parameter.

Appendix 4.1

Megawat-Frequency Control Problem of
Multiarea Electric Energy Systems

The state variable model of the megawatt-frequency control problem of multiarea electric energy systems was developed in (Elgerd and Fosha, 1970, Fosha and Elgerd, 1970). The same model was used in (Geromel and Peres, 1985) for decentralized load-frequency control.

The model development was based on the following two assumptions:

a) For incremental changes in demand power, the two problems, control of real power and frequency, and control of reactive power and voltage, are decoupled and can be considered separately. The megawatt-frequency control problem is the first of the two problems. The frequency deviation is kept within prescribed limits by maintaining control over the real power.

b) The individual electrical connections within an area are so strong at least in comparison to the ties between adjoing areas, that each area may be represented by a single frequency. Therefore, it is assumed that all generators in a single area swing in unision during changes in area load. This characteristic of an area is called coherency (see Figure 4.2)

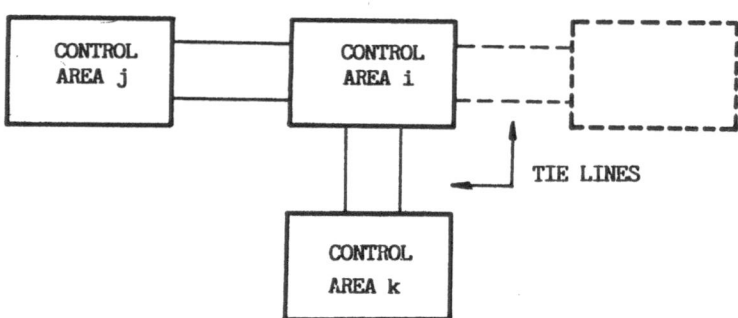

Figure 4.2 Interconnections of individual control areas (coherent areas)

In developing the space model the following, basis equations were used:

i) Power equilibrium equation in per unit for area i:

$$\frac{2H_i}{f^*} \frac{d}{dt} \Delta f_i + D_i \Delta f_i + \Delta P_{tie\ i} = \Delta P_{gi} - \Delta P_{di}$$

where

$H_i = \dfrac{W_{kin}}{P_{ri}}$ inertia constant (seconds)

W_{kin} kinetic energy

$\dfrac{2 W_{kin}}{f^*} \dfrac{d}{dt} \Delta f_i$ increased kinetic energy

f^* nominal fequency

D_i rate at which system load changes with frequency evaluated at nominal frequency f^* (MW/Hz)

$D_i \Delta f_i$ increased load consumption

P_{ri} rated power of area i (MW)

$\Delta P_{tie\ i}$ increased export of power over tie lines

ΔP_{gi} increased generation

ΔP_{di} increased demand

ii) Expression for incremental changes in tie-line power in area i

$$\Delta P_{tie\ i} = \sum_\nu T^*_{i\nu} (\textstyle\int \Delta f_i dt - \int \Delta f_\nu dt)$$

where

$$T^*_{i\nu} \triangleq 2\Pi \frac{|V_i| |V_\nu|}{X_{i\nu} P_{ri}} \cos(\delta^*_i - \delta^*_\nu)$$

$$V_i = |V_i|e^{j\delta i}$$

$$V_\nu = |V_\nu|e^{j\delta i}$$

$X_{i\nu}$ reactance of a lossless line

The total increment in exported power from area i is symbolized in the block diagram form in Figure 4.3.

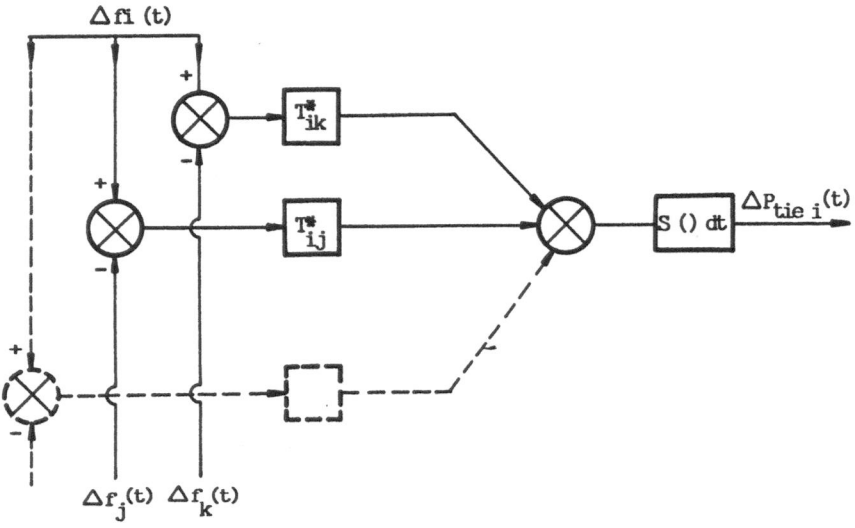

Figure 4.3 Incremental tie-line power out of area i

iii) Incremental generated power equation

$$\frac{d}{dt}\Delta P_g = -\frac{1}{T_t}\Delta P_g + \frac{1}{T_i}\Delta X_{g\nu}$$

$$\frac{d}{dt}\Delta X_{g\nu} = -\frac{1}{T_{g\nu}}\Delta X_{g\nu} - \frac{1}{T_{g\nu}R}\Delta f + \frac{1}{T_{g\nu}}\Delta P_c$$

where

ΔP_g incremental change in generation (pu MW)

$\Delta X_{g\nu}$ incremental change in the governor valve position (pu MW)

R   self-regulation of the generator (Hz/pu MW)

ΔP_c incremental change in the speed changer position (pu MW)

T_t time constant of the turbine

T_{gv} time constant of the governer

The generator system is illustrated in Figure 4.4

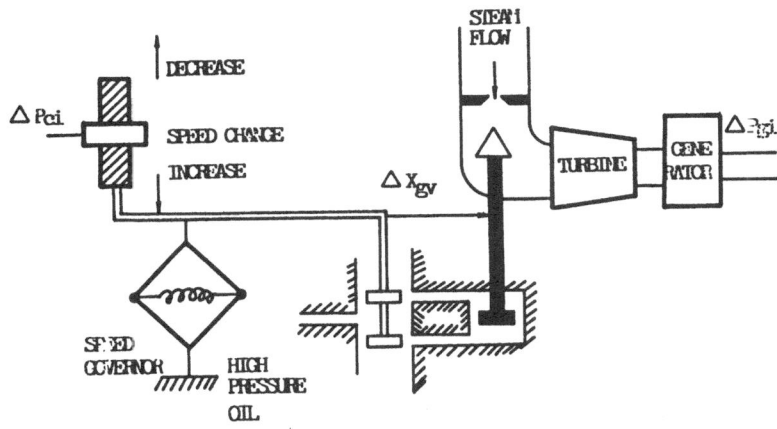

Figure 4.4 Typical turbine control arrangement

Taking the equations for the power equilibrium, the incremental tie-line flow, and the change in generation and position of the speed governer in area i, we have

$$\frac{2H_i}{f^*}\frac{d}{dt}\Delta f_i + D_i\Delta f_i + \sum_v T^*_{iv}\left(\int \Delta f_i dt - \int \Delta f_v dt\right) = \Delta P_{gi} - \Delta P_{di}$$

$$\frac{d}{dt}\Delta P_{gi} = -\frac{1}{T_{ti}}\Delta P_{gi} + \frac{1}{T_{ti}}\Delta X_{gvi}$$

$$\frac{d}{dt}\Delta X_{gvi} = -\frac{1}{T_{gvi}}\Delta X_{gvi} - \frac{1}{T_{gvi}}\frac{\Delta f_i}{R_i} + \frac{1}{T_{gv}}\Delta P_{ci}$$

For a two-area interconnected system the following state and control variables can be defined

$$X \triangleq \begin{bmatrix} \int \Delta P_{tie\,1}\, dt \\ \int \Delta f_1 dt \\ \Delta f_1 \\ \Delta P_{g1} \\ \Delta X_{gv1} \\ \int \Delta f_2 dt \\ \Delta f_2 \\ \Delta P_{g2} \\ \Delta X_{gv2} \end{bmatrix} \qquad u \triangleq \begin{bmatrix} \Delta P_{c1} \\ \Delta P_{c2} \end{bmatrix}$$

The corresponding system matrices A and B are defined with

$$A \triangleq \begin{bmatrix}
0 & T_{12}^* & 0 & 0 & 0 & -T_{12}^* & 0 & 0 & 0 \\
0 & 0 & 1 & 0 & 0 & 0 & 0 & 0 & 0 \\
0 & -\dfrac{f^* T_{12}^*}{2H_1} & \dfrac{f^* D_1}{2H_1} & \dfrac{f^*}{2H_1} & 0 & \dfrac{f^* T_{12}}{2H_1} & 0 & 0 & 0 \\
0 & 0 & 0 & -\dfrac{1}{T_{t1}} & \dfrac{1}{T_{t1}} & 0 & 0 & 0 & 0 \\
0 & 0 & -\dfrac{1}{T_{gv1} R_1} & 0 & -\dfrac{1}{T_{gv1}} & 0 & 0 & 0 & 0 \\
0 & 0 & 0 & 0 & 0 & 0 & 1 & 0 & 0 \\
0 & \dfrac{a_{12} f^* T_{12}^*}{2H_2} & 0 & 0 & 0 & \dfrac{a_{12} f^* T_{12}^*}{2H_2} & -\dfrac{f^* D_2}{2H_2} & \dfrac{f^*}{2H_2} & 0 \\
0 & 0 & 0 & 0 & 0 & 0 & 0 & -\dfrac{1}{T_{t2}} & \dfrac{1}{T_{t2}} \\
0 & 0 & 0 & 0 & 0 & 0 & -\dfrac{1}{T_{gv2} R_2} & 0 & -\dfrac{1}{T_{gv2}}
\end{bmatrix}$$

$$B^T \triangleq \begin{bmatrix}
0 & 0 & 0 & 0 & \dfrac{1}{T_{gv1}} & 0 & 0 & 0 & 0 \\
0 & 0 & 0 & 0 & 0 & 0 & 0 & 0 & \dfrac{1}{T_{gv2}}
\end{bmatrix}.$$

Notice that the disturbance distribution matrix (Fosha and Elgerd, 1970) is not included in this model.

The following system data were used for numerical calculation

$P_{r1} = P_{r2} = 200$ MW

$H_1 = H_2 = 5$ sec.

$D_1 = D_2 = 8.33 \times 10^{-3}$ pu MW/Hz

$T_{t1} = T_{t2} = 0.3$ sec.

$T_{gv\ 1} = T_{gv\ 2} = 0.03$ sec.

$R_1 = R_2 = 2.4$ Hz/pu MW

$P_{tie\ max} = 200$ MW

$\delta^*_1 - \delta^*_2 = 30$ degree

$T_{12}^* = 0.545$ pu MW

$\Delta P_{d1} = 0.01$ pu MW

RECURSIVE APPROACH TO FINITE TIME SINGULARLY PERTURBED AND WEAKLY COUPLED LINEAR CONTROL SYSTEMS

In this chapter we study the main equation of the finite time optimal linear-quadratic control problem, namely, the differential Riccati equation, for both singularly perturbed and weakly coupled systems. A unique approach to the solution of these Riccati equations is developed by performing the block diagonalization of the corresponding Hamiltonian matrices.

5.1 Reduced-Order Recursive Solution of the Singularly Perturbed Differential Riccati Equation

A differential Riccati equation of a singularly perturbed system, (Kokotović and Khalil, 1986), is given by

$$-\dot{P}(t) = P(t)A + A^T P(t) + Q - P(t)SP(t), \qquad P(T) = F \qquad (5.1)$$

where

$$A = \begin{bmatrix} A_1 & A_2 \\ \dfrac{A_3}{\varepsilon} & \dfrac{A_4}{\varepsilon} \end{bmatrix} \qquad Q = \begin{bmatrix} Q_1 & Q_2 \\ Q_2^T & Q_3 \end{bmatrix}, \qquad Q \geq 0$$

$$S = BR^{-1}B^T = \begin{bmatrix} S_1 & \dfrac{Z}{\varepsilon} \\ \dfrac{Z^T}{\varepsilon} & \dfrac{S_2}{\varepsilon^2} \end{bmatrix}, \qquad F = \begin{bmatrix} F_1 & \varepsilon F_2 \\ \varepsilon F_2^{\,T} & \varepsilon F_3 \end{bmatrix}, \qquad R > 0$$

are $n \times n$ constant matrices and ε is a small positive parameter. The presence of a small parameter ε makes this problem numerically ill-defined producing a so called stiff numerical problem (huge slope at terminal time), (Miranker, 1981). In order to overcome this difficulty a Taylor series expansion approach, with respect to a small parameter ε, has been taken in (Yackel and Kokotović, 1973) leading to a family of well-defined reduced-order problems. However, the Taylor series expansion method is not recursive in its application. When one is interested in a high degree of accuracy, or when ε is not very small the size of computations required can be considerable. In such cases, the advantage of using the series expansion method (the important theoretical tool) is questionable from the numerical point of view, and sometimes (see example in Section 5.2) that method is almost not applicable.

In this section we will exploit the known Hamiltonian from of the solution of the Riccati equation (Kwakernaak and Sivan, 1972), and a nonsingular transformation due to Chang (1972) in order to obtain an efficient recursive numerical method for solving (5.1). The Chang transformation is used to block diagonalize the Hamiltonian, so that the required solution is obtained in terms of reduced-order problems. In addition, an efficient Newton type algorithm (with the quadratic rate of convergence, that is, $O(\varepsilon^{2k})$ - where k is a number of iterations) is developed for solving algebraic equations comprising the Chang transformation.

The solution of (5.1) can be sought in the form

$$P(t) = M(t)N^{-1}(t) \qquad\qquad (5.2)$$

where matrices $M(t)$ and $N(t)$ satisfy a system of linear equations (Kwakernaak and Sivan, 1972)

$$\dot{M}(t) = -A^T M(t) - QN(t), \qquad M(T) = F \qquad\qquad (5.3)$$

$$\dot{N}(t) = -SM(t) + AN(t), \qquad N(T) = I \qquad\qquad (5.4)$$

and N(t) is assumed to be nonsingular for ∀t, t < T. This approach is considered as the most efficient numerical method for the solution of the differential Riccati equation (Kenney and Leipnik, 1985), where the invertibility problem of N(t) is solved by performing a reinitialization along the path t_0 < t < T when ever N(t) is close to being singular.

Knowing the nature of the solution of (5.1), which is properly scaled as (Kokotović and Khalil, 1986; Yackel and Kokotović, 1972)

$$P(t) = \begin{bmatrix} P_1(t) & \varepsilon P_2(t) \\ \varepsilon P_2^T(t) & \varepsilon P_3(t) \end{bmatrix}, \quad P(T) = F = \begin{bmatrix} F_1 & \varepsilon F_2 \\ \varepsilon F_2^T & \varepsilon F_3 \end{bmatrix} \tag{5.5}$$

where dim $P_1 = n_1 \times n_1$, dim $P_3 = n_2 \times n_2$, $n = n_1 + n_2$ (n_1 - slow variables, n_2 - fast variables, Kokotović and Khalil, 1986) we introduce compatible partitions of M(t) and N(t) matrices

$$M(t) = \begin{bmatrix} M_1(t) & M_2(t) \\ M_3(t) & M_4(t) \end{bmatrix}, \quad N(t) = \begin{bmatrix} N_1(t) & N_2(t) \\ N_3(t) & N_4(t) \end{bmatrix} \tag{5.6}$$

The invertibility of N(t) for every t, $t_0 \le t < T$, plays an important role in the proposed method. The condition under which N(t) is an invertible matrix is stated in the following lemma.

Lemma 5.1 If the triple (A, B, $\sqrt{Q}$) is stabilizable-observable then the matrix N(t), with N(T) = I is invertible for any t ϵ (t_0 ,T)

Proof: By using dichotomy transformation introduced in (Wilde and Kokotović, 1972),

$$\begin{bmatrix} M \\ N \end{bmatrix} = \begin{bmatrix} \underline{K} & \underline{P} \\ I & I \end{bmatrix} \begin{bmatrix} \hat{M} \\ \hat{N} \end{bmatrix} \tag{5.7}$$

$$\begin{bmatrix} \hat{M} \\ \hat{N} \end{bmatrix} = \begin{bmatrix} (\underline{K} - \underline{P})^{-1} & -(\underline{K} - \underline{P})^{-1}\underline{P} \\ -(\underline{K} - \underline{P})^{-1} & I + (\underline{K} - \underline{P})^{-1}\underline{P} \end{bmatrix} \begin{bmatrix} M \\ N \end{bmatrix} \tag{5.8}$$

where $\underline{P}$ and $\underline{K}$ are unique positive definite and negative definite solutions of the algebraic Riccati equation corresponding to (5.1), the system (5.3)-(5.4) can be transformed in

$$
\begin{bmatrix} \dot{\hat{M}} \\ \dot{\hat{N}} \end{bmatrix} = \begin{bmatrix} A - S\underline{K} & 0 \\ 0 & A - S\underline{P} \end{bmatrix} \begin{bmatrix} \hat{M} \\ \hat{N} \end{bmatrix}
\tag{5.9}
$$

with terminal conditions

$$\hat{M}(T) = (\underline{K} - \underline{P})^{-1}(F - \underline{P})$$

$$\hat{N}(T) = I + (\underline{K} - \underline{P})^{-1}(F - \underline{P}) = I + \hat{M}(T)$$

It is known that $(A - S\underline{K})$ is an unstable matrix and that matrix $(A - S\underline{P})$ is stable (Wilde and Kokotović, 1972). The solution of (5.9) is given by

$$\hat{M}(t) = e^{(A-S\underline{K})(t-T)}\hat{M}(T)$$

$$\hat{N}(t) = e^{(A-S\underline{P})(t-T)}N(T)
\tag{5.10}$$

Using (5.7)-(5.10) it can be easily shown that

$$N(t) = e^{(A-S\underline{P})(t-T)}\left[I + \left(I - e^{S(\underline{P}-\underline{K})(t-T)}(\underline{K} - \underline{P})^{-1}(\underline{P} - F)\right]N(T)\right.$$

that is

$$N(t) = \Phi(t-T)N(T)
\tag{5.11}$$

with obvious definition of $\Phi(t-T)$. Since $\Phi(t-T)$ plays the role of the transition matrix of $N(t)$, and by very well known facts is nonsingular, the regularity of $N(t)$ is determined by $N(T)$ only. Thus having chosen $N(T)$ as an identity will assure the nonsingularity of $N(t)$ for any $t < T$, and prove the given lemma.

Partitioning (5.3) and (5.4), according to (5.6), will reveal a decoupled structure, that is, equations for M_1, M_3, N_1 and N_3 are independent of equations for M_2, M_4, N_2 and N_4 and vice versa. Introducing a notation

$$U = \begin{bmatrix} M_1 \\ N_1 \end{bmatrix}, \quad \varepsilon V = \begin{bmatrix} M_3 \\ \varepsilon N_3 \end{bmatrix}, \quad X = \begin{bmatrix} M_2 \\ N_2 \end{bmatrix}, \quad \varepsilon Y = \begin{bmatrix} M_4 \\ \varepsilon N_4 \end{bmatrix} \tag{5.12}$$

$$T_1 = \begin{bmatrix} -A_1^T & -Q_1 \\ -S_1 & A_1 \end{bmatrix}, \quad T_2 = \begin{bmatrix} -A_3^T & -Q_2 \\ -Z & A_2 \end{bmatrix} \tag{5.13}$$

$$T_3 = \begin{bmatrix} -A_2^T & -Q_2^T \\ -Z^T & A_3 \end{bmatrix}, \quad T_4 = \begin{bmatrix} -A_4^T & -Q_3 \\ -S_2 & A_4 \end{bmatrix}$$

and after doing some algebra, we get two systems of singularly perturbed matrix equations

$$\dot{U} = T_1 U + T_2 V \qquad U(T) = \begin{bmatrix} F_1 \\ I \end{bmatrix}$$

$$\tag{5.14}$$

$$\varepsilon \dot{V} = T_3 U + T_4 V \qquad V(T) = \begin{bmatrix} F_2 \\ 0 \end{bmatrix}$$

$$\dot{X} = T_1 X + T_2 Y \qquad X(T) = \begin{bmatrix} \varepsilon F_2 \\ 0 \end{bmatrix}$$

$$\tag{5.15}$$

$$\varepsilon \dot{Y} = T_3 X + T_4 Y \qquad Y(T) = \begin{bmatrix} F_3 \\ I \end{bmatrix}$$

Note that these two systems have exactly the same form and they differ in terminal conditions only. From this point we will proceed by applying the Chang transform to (5.14) and (5.15). This transformation is defined by (Chang, 1972)

$$J = \begin{bmatrix} I - \varepsilon HL & -\varepsilon H \\ L & I \end{bmatrix} \tag{5.16}$$

and

$$J^{-1} = \begin{bmatrix} I & \varepsilon H \\ -L & I - \varepsilon LH \end{bmatrix} \tag{5.17}$$

where L and H satisfy

$$T_4 L - T_3 - \varepsilon L(T_1 - T_2 L) = 0 \tag{5.18}$$

$$-H(T_4 + \varepsilon LT_2) + T_2 + \varepsilon(T_1 - T_2 L)H = 0 \tag{5.19}$$

Applied to (5.14) and (5.15) it will produce

$$\dot{\hat{U}} = (T_1 - T_2 L)\hat{U} \qquad\qquad \hat{U}(T) = (I - \varepsilon HL)U(T) - \varepsilon HV(T) \tag{5.20}$$

$$\varepsilon\dot{\hat{V}} = (T_4 + \varepsilon LT_2)\hat{V} \qquad\qquad \hat{V}(T) = LU(T) + V(T) \tag{5.21}$$

$$\dot{\hat{X}} = (T_1 - T_2 L)\hat{X} \qquad\qquad \hat{X}(T) = (I - \varepsilon HL)X(T) - \varepsilon HY(T) \tag{5.22}$$

$$\varepsilon\dot{\hat{Y}} = (T_4 + \varepsilon LT_2)\hat{Y} \qquad\qquad \hat{Y}(T) = LX(T) + Y(T) \tag{5.23}$$

Solutions of (5.20)-(5.23) are given by

$$\hat{U}(t) = e^{(T_1 - T_2 L)(t-T)}\,\hat{U}(T) \tag{5.24}$$

$$\hat{V}(t) = e^{\frac{1}{\varepsilon}(T_4 + \varepsilon LT_2)(t-T)}\,\hat{V}(T) \tag{5.25}$$

$$\hat{X}(t) = e^{(T_1 - T_2 L)(t-T)}\,\hat{X}(T) \tag{5.26}$$

$$\hat{Y}(t) = e^{\frac{1}{\varepsilon}(T_4 + \varepsilon LT_2)(t-T)}\,\hat{Y}(T) \tag{5.27}$$

so that in the original coordinates we have

$$U(t) = e^{(T_1 - T_2 L)(t-T)}\,\hat{U}(T) + \varepsilon H\, e^{\frac{1}{\varepsilon}(T_4 + \varepsilon LT_2)(t-T)}\,\hat{V}(T) \tag{5.28}$$

$$V(t) = -L \ e^{(T_1 - T_2 L)(t-T)} \ \hat{U}(T) + (I - \varepsilon LH) e^{\frac{1}{\varepsilon}(T_4 + \varepsilon LT_2)(t-T)} \hat{V}(T) \qquad (5.29)$$

$$X(t) = e^{(T_1 - T_2 L)(t-T)} \ \hat{X}(T) + \varepsilon H \ e^{\frac{1}{\varepsilon}(T_4 + \varepsilon LT_2)(t-T)} \ \hat{Y}(T) \qquad (5.30)$$

$$Y(t) = -L \ e^{(T_1 - T_2 L)(t-T)} \ \hat{X}(T) + (I - \varepsilon LH) e^{\frac{1}{\varepsilon}(T_4 + \varepsilon LT_2)(t-T)} \ \hat{Y}(T) \qquad (5.31)$$

Partitioning (5.28)-(5.31) according to (5.12) will produce all components of matrices $M(t)$ and $N(t)$, that is

$$\begin{bmatrix} M_1(t) \\ N_1(t) \end{bmatrix} = \begin{bmatrix} U_1(t) \\ U_2(t) \end{bmatrix} = U(t), \qquad \begin{bmatrix} M_2(t) \\ N_2(t) \end{bmatrix} = \begin{bmatrix} X_1(t) \\ X_2(t) \end{bmatrix} = X(t)$$

$$\begin{bmatrix} \frac{1}{\varepsilon} M_3(t) \\ N_3(t) \end{bmatrix} = \begin{bmatrix} V_1(t) \\ V_2(t) \end{bmatrix} = V(t), \qquad \begin{bmatrix} \frac{1}{\varepsilon} M_4(t) \\ N_2(t) \end{bmatrix} = \begin{bmatrix} Y_1(t) \\ Y_2(t) \end{bmatrix} = Y(t)$$

so that the required solution of (5.1) is given by

$$P(t) = \begin{bmatrix} U_1(t) & X_1(t) \\ \varepsilon V_1(t) & \varepsilon Y_1(t) \end{bmatrix} \begin{bmatrix} U_2(t) & X_2(t) \\ V_2(t) & Y_2(t) \end{bmatrix}^{-1} \qquad (5.32)$$

Thus, in order to get the numerical solution of (5.1), that is $P(t)$, which has dimensions $n \times n = (n_1 + n_2) \times (n_1 + n_2)$, we have to solve two simple algebraic equations (5.18) and (5.19) of dimensions of $(2n_2 \times 2n_1)$ and $(2n_1 \times 2n_2)$ respectively. The existing numerical algorithms for solving (5.18) and (5.19) can be found in (Gajić, 1986; Kokotović, Allemong, Winkelman and Chow, 1980). Then, two exponential forms $\exp\left[(T_1 - T_2 L)(t-T)\right]$ and $\exp\left[1/\varepsilon (T_4 + \varepsilon LT_2)(t-T)\right]$, have to be transformed in the matrix forms by using some of the well known approaches (Molen and Von Loan, 1978). Finally, the inversion of the matrix $N(t)$ has to be performed.

The algebraic equations: (5.18) which is a weakly nonlinear equation and (5.19) a linear Lyapunov type equation, play the crucial role in the developed method and a very important role in the linear theory of singular perturbations (Kokotović and Khalil, 1986). The existing methods for solving (5.18) and (5.19) are recursive type algorithms with a rate of convergence of $O(\varepsilon^k)$, where k represents the number of iterations (Gajić, 1986; Kokotović, Allemong, Winkelman and Chow, 1980). In this section a new method for solving (5.18) and (5.19) with a quadratic rate of convergence, that is $O(\varepsilon^{2k})$, will be developed. This method is based on the Newton type recursive scheme. It is a very well known fact that the Newton method converges quadratically in the neighbourhood of the sought solution and that its main problem is in the choice of the initial guess. For the algebraic equation (5.18) the initial guess is easily obtained with the accuracy of $O(\varepsilon)$, by setting $\varepsilon = 0$ in that equation, that is

$$L^{(0)} = T_4^{-1}T_3 = L + O(\varepsilon) \qquad (5.33)$$

Thus, the Newton sequence will be $O(\varepsilon^2)$, $O(\varepsilon^4)$, $O(\varepsilon^8)$,...., $O(\varepsilon^{2k})$ close to the exact solution, respectively in each iteration.

The Newton type algorithm of (5.18), can be constructed by setting $L^{(i+1)} = L^{(i)} + \Delta L^{(i)}$ and neglecting $O(\Delta L)^2$ terms. This will produce a Lyapunov type equation of the form

$$D_1^{(i)}L^{(i+1)} + L^{(i+1)}D_2^{(i)} = Q^{(i)} \qquad (5.34)$$

where

$$D_1^{(i)} = T_4 + \varepsilon L^{(i)}T_2, \qquad\qquad D_2^{(i)} = -\varepsilon(T_1 - T_2 L^{(i)})$$

$$Q^{(i)} = T_3 + \varepsilon L^{(i)}T_2 L^{(i)}, \quad i = 0, 1, 2 \dots$$

with the initial condition given by (5.33).

Having found the solution of (5.18), up to the required degree of accuracy, one can get the solution of (5.19) by solving directly a Lyapunov equation of the form

$$M^{(i)}D_1^{(i)} + D_2^{(i)}M^{(i)} = T_2 \qquad (5.35)$$

which implies $M^{(i)} = M + O(\varepsilon^{2^i})$

Note that the existence of the solutions of (5.18) and (5.19) is guaranteed by the nonsingularity of T_4. The sufficient condition for the convergence of the algorithm (5.34) is given by (Belanger and McGillivray 1976)

$$\| \, \Delta L^{(i)} \, \| \leq \| \, Q^{(i)} \, \| = \| \, T_3 + \varepsilon L^{(i)} T_2 L^{(i)} \, \| \tag{5.36}$$

which is almost always satisfied, except for some special cases, for example $T_3 \approx 0$ and $T_2 \approx 0$, which corresponds to a system already in a block diagonal form.

One has to point out, that in the contrary to previously used algorithms for solving (5.18)-(5.19) (Gajić, 1986; Kokotović, Allemong, Winkelman and Chow, 1980), which require recursive solution of linear equations, in the proposed method one is faced with the recursive solution of Lyapunov equations. Thus, for the price of speeding up the convergence from $O(\varepsilon^k)$ to $O(\varepsilon^{2^i})$ slightly more computations have to be performed per iteration. However, the size of computations required is of the same order, that is of $O(n^3)$ for both the solution of the Lyapunov and solution of linear equations, so that the comparison of the rate of convergence of these two algorithms plays the dominant role. In order to demonstrate the efficiency of the proposed algorithm, we have run a fifth order example.

Example 5.1

Matrices T_1, T_2, T_3 and T_4 are chosen randomly (standard deviation equals to 1, and mean value equals to zero) such that T_4 is the invertible matrix. The simulation results for different values of a small parameter are given in Table 5.1. It can be seen that the Newton method is much more powerful than the successive approximation recursive scheme (Gajić, 1986; Kokotović, Allemong, Winkelman and Chow, 1980). In the second table we have shown the propogation of the error per iteration when $\varepsilon = 0.2$ for the Newton method.

T_1

-2.014	-0.058	0.499	0.585	1.372
1.366	-0.805	0.320	0.548	0.950
-0.952	0.747	0.984	-1.816	-1.563
-1.241	0.758	-1.126	0.497	-0.131
0.663	-0.021	-0.640	-0.296	1.375

T_2

-1.796	-0.009	-0.840	1.819	0.794
0.158	0.467	1.324	-0.123	0.629
-0.433	0.248	-1.181	-1.426	0.297
-1.599	0.269	-0.133	-0.845	-0.769
1.967	-0.565	0.776	1.419	-0.450

T_3

-1.496	-0.666	0.699	1.262	-0.731
1.343	0.563	0.812	-1.300	-0.616
-0.521	-0.962	-0.141	-1.159	0.939
1.071	-0.943	0.017	0.696	1.295
1.397	-1.436	0.843	-1.488	0.524

T_4

-1.367	-0.885	-0.506	-1.174	1.435
0.133	1.319	1.244	0.892	-1.221
-0.296	1.333	1.002	-0.927	-0.794
0.780	1.358	0.607	-0.511	0.671
-0.999	0.914	-1.320	-0.556	-1.135

The Hamiltonian method developed in this section will be used for the numerical solution of the singularly perturbed matrix differential Riccati equation. Since the matrices $M(t)$ and $N(t)$ contain unstable modes of the Hamiltonian also, (Kwakernaak and Sivan, 1972), then even though the product $M(t)$ and $N^{-1}(t)$ tends to a constant as $t \to \infty$, the inversion of the nonsingular matrix $N(t)$, which contains huge elements, will hurt the accuracy.

ε	number of required iterations such that $\| L^{(i+1)} - L^{(i)} \|_\infty < 10^{-7}$	
	Newton method	Successive approximations
0.3	6	*
0.2	5	*
0.1	4	*
0.04	4	19
0.02	4	11
0.01	3	7
0.001	2	4

*does not converge

Table 5.1. Dependence of the number of iterations on ε

$\varepsilon = 0.2$	
i	$\| L^{(i+1)} - L^{(i)} \|_\infty < 10^{-7}$
1	2.40745×10^0
2	7.80653×10^{-1}
3	4.21800×10^{-2}
4	0.88748×10^{-4}
5	0.17808×10^{-8}

Table 5.2 Propagation of the error per iteration for a constant value of
ε for the Newton method

The reinitialization version of the Hamiltonian approach, which leads to the known Kalman-Englar method (Kwakernaak and Sivan, 1972), is considered as the most efficient numerical method for the solution of the general matrix differential Riccati equation. The reinitialization technique applied to the obtained results will modify formulas (5.3), (5.14)-(5.15) respectively in

$$M(k\Delta t) = P(k\Delta t) \tag{5.37}$$

$$U(k\Delta t) = \begin{bmatrix} P_1(k\Delta t) \\ I \end{bmatrix}, \quad V(k\Delta t) = \begin{bmatrix} P_2^T(k\Delta t) \\ 0 \end{bmatrix} \tag{5.38}$$

$$X(k\Delta t) = \begin{bmatrix} \varepsilon P_2(k\Delta t) \\ 0 \end{bmatrix}, \quad Y(k\Delta t) = \begin{bmatrix} P_3(k\Delta t) \\ I \end{bmatrix} \tag{5.39}$$

where k represents the number of steps and Δt is an integration step. This will introduce slight modifications in formulas (5.20)-(5.31), namely, instead of the final time T a discrete time $k\Delta t$ has to be used. These changes can be implemented very easily from the programming point of view.

The recursive method for the numerical solution for the singularly perturbed Riccati differential equation proposed in this section is very important in two cases: a) ε is not very small, b) high order of accuracy is required. The first case represents one of the main problems in the modern numerical analysis of the singularly perturbed problems. It was pointed by P. Hemker (1983), that "numerical analysis of singular perturbation problems mainly concentrates on the following question: how to find a numerical approximation to the solution for small as well as intermediate values of ε, where no short asymptotic expansion is available. Or, more general, how to construct a single numerical method that can be applied both in the case of extremely small ε and for larger values of ε, when one wouldn't consider the problem as singularly perturbed any longer". Results reported in this section resolve that problem in the case of the singularly perturbed Riccati differential equation.

5.2 Case Study: The Synchronous Machine Connected to an Infinite Bus

The recursive solution of the differential matrix Riccati equation of singularly perturbed systems is demonstrated of the seventh order model of the synchronous machine connected to an infinite bus (Kokotović, Allemong, Winkelman and Chow, 1980) - see Appendix 5.1. The system matrix A is given by

$$
A = \begin{bmatrix}
-0.58 & 0 & 0 & -0.27 & 0 & 0.2 & 0 \\
0 & -1. & 0 & 0 & 0 & 1. & 0 \\
0 & 0 & -5. & 2.1 & 0 & 0 & 0 \\
0 & 0 & 0 & 0 & 337. & 0 & 0 \\
-0.14 & 0 & 0.14 & -0.2 & -0.28 & 0 & 0 \\
0 & 0 & 0 & 0 & 0 & 0.08 & 2. \\
-173. & 66.7 & -116. & 40.9 & 0 & -66.7 & -16.7
\end{bmatrix}
$$

Remaining matrices are chosen as $Q = I$, $F = 0$, S_1, S_2 and Z have all entries equal to 1. The eigenvalues of A are -8.53 ± j8.22, -3.93, -0.326 ± j0.56, -0.86 ± j8.37. Two fast and five slow variables are separated by the choice of the small singular perturbation parameter $\varepsilon = 0.4$ (roughly the ratio of 3.93 and 8.53). Simulation results for the element $P_{11}(t)$ are given in Table 5.3.

time = t	0.1	0.5	1.0
P_{11} (exact)	1.9699	6.6483	9.6600
P_{11} (12)	"	"	"
P_{11} (11)	"	"	9.6599
P_{11} (10)	1.9698	"	9.6601
P_{11} (9)	1.9700	6.6484	9.6598
P_{11} (8)	1.9696	6.6482	9.6602
P_{11} (7)	1.9703	6.6487	9.6603
P_{11} (6)	1.9694	6.6471	9.6572
P_{11} (5)	1.9703	6.6500	9.6671
P_{11} (4)	1.9720	6.6496	9.6477
P_{11} (3)	1.9537	6.6488	9.6991
P_{11} (2)	2.0603	6.6520	9.5417
P_{11} (1)	1.9847	6.7926	9.8624
P_{11} (0)	1.9742	7.0256	10.4610

Table 5.3 Simulation results for the element $P_{11}(t)$

It can be seen that in order to get the accuracy of four decimal digits it takes 12 iterations (the successive approximation method was used for solving algebraic equations composing the Chang transformation - in order to be able to compare the proposed recursive scheme with the power series expansion method, since both methods are producing the same order of accuracy). This result is expected since $O(0,4^{12}) \approx 10^{-5}$. That means if the power series expansion method had been used, in order to get the same accuracy, it would have required 12 terms, that is (Yackel and Kokotović, 1973)

$$P(t, \varepsilon) = \sum_{m=0}^{11} \frac{\varepsilon^m}{m!} \left\{ P_s^{(m)}(t) + P_f^{(m)}(\tau) \right\} + O(\varepsilon^{12}), \ \tau = \frac{t - T}{\varepsilon}$$

where

$$P_s^{(m)}(t) = \begin{bmatrix} P_{1s}^{(m)}(t) & \varepsilon P_{2s}^{(m)}(t) \\ \varepsilon P_{2s}^{(m)T}(t) & \varepsilon P_{3s}^{(m)}(t) \end{bmatrix}, \quad P_f^{(m)}(\tau) = \begin{bmatrix} P_{1f}^{(m)}(\tau) & \varepsilon P_{2f}(\tau) \\ \varepsilon P_{2f}^{(m)T}(\tau) & \varepsilon P_{3f}^{(m)}(\tau) \end{bmatrix}$$

It is shown in (Yackel and Kokotović, 1973), (pp. 21, formula 32) that the right hand sides of differential equations for $P_{1f}^{(1)}(\tau)$, $P_{2f}^{(1)}(\tau)$ and $P_{3f}^{(1)}(\tau)$ contain respectively 7, 23 and 22 terms, each consisting of a product of two or three matrices. Thus, the size of computations required for only an $O(\varepsilon^2)$ accuracy is already enormous. The complexity of the right hand side of differential equations for $P_f^{(m)}(\tau)$ grows extremely quickly with the increase of m so that this nice theoretical method is not convenient for the practical computations. For an $O(\varepsilon^{12})$ accuracy, the right hand sides of the differential equations for the power series expansion method will contain hundreds or even thousands of terms, and this example can not be efficiently solved by using that method.

5.3 Reduced-Order Recursive Solution of the Riccati Differential Equation of Weakly Coupled Systems

The recursive approach to weakly coupled systems, bases on the fixed point iterations, is developed in (Gajić and Rayavarupu, 1989, Gajić and Shen, 1989a, Harkara, Petkovski and Gajić, 1989, Petrović and Gajić, 1988, Shen and Gajić, 1989a,b,c, Su and Gajić, 1989). It has been shown that the recursive methods are particularly useful when the coupling parameter ε is not extremely small and/or when any desired order of accuracy is required, namely, $O(\varepsilon^k)$, where k = 2, 3, 4, ...

The recursive methods of (Gajić and Rayavarupu, 1989, Gajic and Shen, 1989a, Harkara, Petkovski and Gajić, 1989, Petrović and Gajić 1988, Shen and Gajić, 1989a, 1989b, 1989c) are based on the fixed point theory applied to the corresponding algebraic equations, so that these results are applicable to the steady state control problems only.

In this section we will study the finite time optimal closed loop control problem of weakly coupled systems by following results from (Su and Gajić, 1989). The solution of this problem is given in terms of the Riccati differential equation, which makes it more challenging for research.

The recursive reduced-order solution will be obtained by exploiting the transformation introduced in (Gajić and Shen, 1989a)- see Section 2.4, which will block diagonalize the Hamiltonian form of the solution for the optimal linear-quadratic control problem. Completely decoupled sets of reduced-order differential equations are obtained. The convergence to the optimal solution is pretty rapid, due to the fact that the algorithms derived in (Gajić and Shen, 1989a) has the rate of convergence of at least of $O(\varepsilon^2)$. This produces a lot of savings in the size of computations required. In addition, the proposed method is very suitable for the parallel computations.

Consider a linear weakly coupled system

$$\dot{x}_1 = A_1 x_1 + \varepsilon A_2 x_2 + B_1 u_1 + \varepsilon B_2 u_2 , \qquad x_1(t_0) = x_{10}$$

$$\tag{5.40}$$

$$\dot{x}_2 = \varepsilon A_3 x_1 + A_4 x_2 + \varepsilon B_3 u_1 + B_4 u_2 , \qquad x_2(t_0) = x_{20}$$

with

$$z = \begin{bmatrix} z_1 \\ z_2 \end{bmatrix} = D \begin{bmatrix} x_1 \\ x_2 \end{bmatrix} = \begin{bmatrix} D_1 & \varepsilon D_2 \\ \varepsilon D_3 & D_4 \end{bmatrix} \begin{bmatrix} x_1 \\ x_2 \end{bmatrix} \tag{5.41}$$

where $x_i \in \mathbf{R}^{n_i}$, $u_i \in \mathbf{R}^{m_i}$, $z_i \in \mathbf{R}^{r_i}$, $i = 1, 2$ are state, control and output variables respectively. The system matrices are of appropriate dimensions and, in general, they are bounded functions of a small coupling parameter ε, (Gajić and Rayavarupu, 1989, Harkara, Petkovski and Gajić, 1989, Petrović and Gajić, 1988). In this section we will assume that all given matrices are constant.

With (5.40)-(5.41) consider the performance criterion

$$J = \frac{1}{2} \int_{t_0}^{T} \left\{ \begin{bmatrix} x_1 \\ x_2 \end{bmatrix}^T D^T D \begin{bmatrix} x_1 \\ x_2 \end{bmatrix} + \begin{bmatrix} u_1 \\ u_2 \end{bmatrix}^T R \begin{bmatrix} u_1 \\ u_2 \end{bmatrix} \right\} dt + \frac{1}{2} \begin{bmatrix} x_1(T) \\ x_2(T) \end{bmatrix}^T F \begin{bmatrix} x_1(T) \\ x_2(T) \end{bmatrix} \tag{5.42}$$

with positive definite R and positive semi-definite F, which has to be minimized. It is assumed that matrices F and R have the weakly coupled structure, that is

$$F = \begin{bmatrix} F_1 & \varepsilon F_2 \\ \varepsilon F_2^T & F_3 \end{bmatrix}, \qquad R = \begin{bmatrix} R_1 & 0 \\ 0 & R_2 \end{bmatrix} \qquad (5.43)$$

The optimal closed-loop control law has the very well known form (Kwakernaak and Sivan, 1972)

$$u = \begin{bmatrix} u_1 \\ u_2 \end{bmatrix} = R^{-1} \begin{bmatrix} B_1 & \varepsilon B_2 \\ \varepsilon B_3 & B_4 \end{bmatrix}^T P \begin{bmatrix} x_1 \\ x_2 \end{bmatrix} = R^{-1}B^T P x \qquad (5.44)$$

where P satisfies the differential Riccati equation given by

$$-\dot{P} = PA + A^T P + D^T D - PSP, \quad P(T) = F \qquad (5.45)$$

with

$$A = \begin{bmatrix} A_1 & \varepsilon A_2 \\ \varepsilon A_3 & A_4 \end{bmatrix}, \qquad S = BR^{-1}B^T = \begin{bmatrix} S_1 & \varepsilon S_2 \\ \varepsilon S_2^T & S_3 \end{bmatrix} \qquad (5.46)$$

Due to a weakly coupled structure of all coefficients in (5.45), the solution of that equation has the form

$$P = \begin{bmatrix} P_1 & \varepsilon P_2 \\ \varepsilon P_2^T & P_3 \end{bmatrix} \qquad (5.47)$$

In this section we will exploit the Hamiltonian form of the solution of the Riccati differential equation and a nonsingular transformation introduced in (Gajić and Shen, 1989a) in order to obtain an efficient recursive method for solving (5.45).

The solution of (5.45) can be sought in the form

$$P(t) = M(t)N^{-1}(t) \qquad (5.48)$$

where matrices M(t) and N(t) satisfy a system of linear equations (Kwakernaak and Sivan, 1972)

$$\dot{M} = -A^T M(t) - D^T DN(t), \qquad M(T) = F \qquad (5.49)$$

$$\dot{N}(t) = -SM(t) + AN(t), \qquad N(T) = I \qquad (5.50)$$

The Lemma 5.1, proved in Section 5.2, guarantees the existence of the invertible solution for $N(t)$.

Knowing the nature of the solution of (5.45) we introduce compatible partitions of $M(t)$ and $N(t)$ matrices as

$$M(t) = \begin{bmatrix} M_1(t) & \varepsilon M_2(t) \\ \varepsilon M_3(t) & M_4(t) \end{bmatrix}, \qquad N(t) = \begin{bmatrix} N_1(t) & \varepsilon N_2(t) \\ \varepsilon N_3(t) & N_4(t) \end{bmatrix} \qquad (5.51)$$

Partitions of (5.49) and (5.50), according to (5.51), will reveal a decoupled structure, that is, M_1, M_3, N_1, and N_3 are independent of equations for M_2, M_4, N_2, and N_4 and vice versa. Let us introduce a notation

$$U = \begin{bmatrix} M_1 \\ N_1 \end{bmatrix}, \quad V = \begin{bmatrix} \varepsilon M_3 \\ \varepsilon N_3 \end{bmatrix}, \quad X = \begin{bmatrix} \varepsilon M_2 \\ \varepsilon N_2 \end{bmatrix}, \quad Y = \begin{bmatrix} M_4 \\ N_4 \end{bmatrix} \qquad (5.52)$$

and

$$T_1 = \begin{bmatrix} -A_1^T & -Q_1 \\ -S_1 & A_1 \end{bmatrix}, \qquad T_2 = \begin{bmatrix} -A_3^T & -Q_2 \\ -S_2 & A_2 \end{bmatrix}$$

$$T_3 = \begin{bmatrix} -A_2^T & -Q_2^T \\ -S_2^T & A_3 \end{bmatrix}, \qquad T_4 = \begin{bmatrix} -A_4^T & -Q_3 \\ -S_3 & A_4 \end{bmatrix} \qquad (5.53)$$

where

$$Q_1 = D_1^T D_1 + \varepsilon^2 D_3^T D_3 , \; Q_2 = D_1^T D_2 + D_3^T D_4 ,$$

$$Q_3 = D_4^T D_4 + \varepsilon^2 D_2^T D_2$$

After doing some algebra, we get two independent systems of weakly coupled matrix differential equations

$$\dot{U} = T_1 U + \varepsilon T_2 V$$
$$\dot{V} = \varepsilon T_3 U + T_4 V$$

(5.54)

with terminal conditions

$$U(T) = \begin{bmatrix} F_1 \\ I \end{bmatrix}, \qquad V(T) = \begin{bmatrix} \varepsilon F_2^T \\ 0 \end{bmatrix}$$

(5.55)

and

$$\dot{X} = T_1 X + \varepsilon T_2 Y$$
$$\dot{Y} = \varepsilon T_3 X + T_4 Y$$

(5.56)

with terminal conditions

$$X(T) = \begin{bmatrix} \varepsilon F_2 \\ 0 \end{bmatrix}, \qquad Y(T) = \begin{bmatrix} F_3 \\ I \end{bmatrix}$$

(5.57)

Note that these two systems have exactly the same form and they differ in terminal conditions only. From this point we will proceed by applying the decoupling transformation introduced in (Gajić and Shen, 1989a). This transformation is defined by

$$K = \begin{bmatrix} I & -\varepsilon L \\ \varepsilon H & I-\varepsilon^2 HL \end{bmatrix}, \qquad K^{-1} = \begin{bmatrix} I - \varepsilon^2 LH & \varepsilon L \\ -\varepsilon H & I \end{bmatrix}$$

(5.58)

where L and H satisfy

$$T_1 L + T_2 - LT_4 - \varepsilon^2 LT_3 L = 0$$

(5.59)

$$H(T_1 - \varepsilon^2 LT_3) - (T_4 + \varepsilon^2 T_3 L) H + T_3 = 0$$

(5.60)

Applied to (5.54)-(5.57), it will produce

$$\dot{\hat{U}} = (T_1 - \varepsilon^2 LT_3)\hat{U} , \quad \hat{U}(T) = U(T) - \varepsilon LV(T) \tag{5.61}$$

$$\dot{\hat{V}} = (T_4 + \varepsilon^2 T_3 L)\hat{V} , \quad \hat{V}(T) = \varepsilon HU(T) + (I - \varepsilon^2 HL)V(T) \tag{5.62}$$

and

$$\dot{\hat{X}} = (T_1 - \varepsilon^2 LT_3)\hat{X} , \quad \hat{X}(T) = X(T) - \varepsilon LY(T) \tag{5.63}$$

$$\dot{\hat{Y}} = (T_4 + \varepsilon^2 T_3 L)\hat{Y} , \quad \hat{Y}(T) = \varepsilon HX(T) + (I - \varepsilon^2 HL)Y(T) \tag{5.64}$$

Solutions of (5.61)-(5.64) are given by

$$\hat{U}(t) = e^{(T_1-\varepsilon^2 LT_3)(t-T)}\hat{U}(T) \tag{5.65}$$

$$\hat{V}(t) = e^{(T_4+\varepsilon^2 T_3 L)(t-T)}\hat{V}(T) \tag{5.66}$$

$$\hat{X}(t) = e^{(T_1-\varepsilon^2 LT_3)(t-T)}\hat{X}(T) \tag{5.67}$$

$$\hat{Y}(t) = e^{(T_4+\varepsilon^2 T_3 L)(t-T)}\hat{Y}(T) \tag{5.68}$$

so that in the original coordinates we have

$$U(t) = (I - \varepsilon^2 LH)e^{(T_1-\varepsilon^2 LT_3)(t-T)}\hat{U}(T) + \varepsilon Le^{(T_4+\varepsilon^2 T_3 L)(t-T)}\hat{V}(T) \tag{5.69}$$

$$V(t) = -\varepsilon He^{(T_1-\varepsilon^2 LT_3)(t-T)}\hat{U}(T) + e^{(T_4+\varepsilon^2 T_3 L)(t-T)}\hat{V}(T) \tag{5.70}$$

$$X(t) = (I - \varepsilon^2 LH)e^{(T_1-\varepsilon^2 LT_3)(t-T)}\hat{X}(T) + \varepsilon Le^{(T_4+\varepsilon^2 T_3 L)(t-T)}\hat{Y}(T) \tag{5.71}$$

$$Y(t) = -\varepsilon He^{(T_1-\varepsilon^2 LT_3)(t-T)}\hat{X}(T) + e^{(T_4+\varepsilon^2 T_3 L)(t-T)}\hat{Y}(T) \tag{5.72}$$

Partitioning $U(t)$, $V(t)$, $X(t)$, and $Y(t)$ according to (5.52) will produce all components of matrices $M(t)$ and $N(t)$, that is

$$U(t) = \begin{bmatrix} U_1(t) \\ U_2(t) \end{bmatrix} = \begin{bmatrix} M_1(t) \\ N_1(t) \end{bmatrix}, \quad V(t) = \begin{bmatrix} V_1(t) \\ V_2(t) \end{bmatrix} = \begin{bmatrix} \varepsilon M_3(t) \\ \varepsilon N_3(t) \end{bmatrix}$$

$$\text{(5.73)}$$

$$X(t) = \begin{bmatrix} X_1(t) \\ X_2(t) \end{bmatrix} = \begin{bmatrix} \varepsilon M_2(t) \\ \varepsilon N_2(t) \end{bmatrix}, \quad Y(t) = \begin{bmatrix} Y_1(t) \\ Y_2(t) \end{bmatrix} = \begin{bmatrix} M_4(t) \\ N_4(t) \end{bmatrix}$$

so that the required solution of (5.45) is given by

$$P(t) = \begin{bmatrix} U_1(t) & X_1(t) \\ V_1(t) & Y_1(t) \end{bmatrix} \begin{bmatrix} U_2(t) & X_2(t) \\ V_2(t) & Y_2(t) \end{bmatrix}^{-1}$$

$$\text{(5.74)}$$

Thus, in order to get the solution of (5.45), P(t), which has dimensions nxn = $(n_1 + n_2) \times (n_1 + n_2)$, we have to solve two simple algebraic equations (5.59) and (5.60) of dimensions $(2n_2 \times 2n_1)$ and $(2n_1 \times 2n_2)$ respectively. The efficient numerical algorithm based on the fixed point iterations and the Newton's method for solving (5.59) and (5.60) are discussed in Section 2.4. Then two exponential forms $\exp[(T_1 - \varepsilon^2 LT_3)(t - T)]$ and $\exp[(T_4 + \varepsilon^2 T_3 L)(t - T)]$, have to be transformed in the matrix forms by using scme of the well-known approaches (Molen and Van Loan, 1978). Finally, the inversion of the matrix N(t) has to be performed.

Since the matrices M(t) and N(t) contain unstable models of the Hamiltonian (Kwakernaak and Sivan, 1972), even though the product $M(t)N^{-1}(t)$ tends to a constant as $t \to \infty$ the inversion of the ronsingular matrix N(t), which contains huge elements, will hurt the accuracy.

The reinitialization version of the Hamiltonian approach avoids that problem. The reinitialization technique applied to the prob.em under consideration will modify only terminal conditions in formulas (5.49), (5.55), and (5.57), respectively.

$$M(k\Delta t) = P(k\Delta t) \tag{5.75}$$

$$U(k\Delta t) = \begin{bmatrix} P_1(k\Delta t) \\ I \end{bmatrix}, \quad V(k\Delta t) = \begin{bmatrix} \varepsilon P_2^T(k\Delta t) \\ 0 \end{bmatrix} \tag{5.76}$$

$$X(k\Delta t) = \begin{bmatrix} \varepsilon P_2 (k\Delta t) \\ 0 \end{bmatrix} , \quad Y(k\Delta t) = \begin{bmatrix} P_3 (k\Delta t) \\ I \end{bmatrix} \tag{5.77}$$

where k represents the number of steps and Δt is an integration step.

The transformation matrix K from (5.58) can be easily obtained, with a required accuracy, by using numerical techniques developed in (Gajić and Shen, 1989a) for solving (5.59)-(5.60). They converge with the rate of convergence of at least of $O(\varepsilon^2)$. Thus, after k iterations, one gets the approximation $K^{(k)} = K + O(\varepsilon^{2k})$. The use of $K^{(k)}$ in (5.61)-(5.64) instead of K, will perturb the coefficients of the corresponding systems of linear differential equations by $O(\varepsilon^{2k})$, which implies that the approximate solutions to these differential equations are $O(\varepsilon^{2k})$ close to the exact ones (Kato, 1980). Thus, it is of interest to obtain $K^{(k)}$ with the desired accuracy, which produces the same accuracy in the sought solution.

The recursive reduced-order solution of the differential Riccati equation of weakly coupled systems is demonstrated in the next section where a real world example is considered.

5.4 Case Study: The Distillaton Column Example

A real world problem, a fifth order distillation column (Petkov, Christov and Konstantinov, 1986) is considered.

The problem matrices A and B are given by

$$A = \begin{bmatrix} -0.1.94 & 0.0628 & 0 & 0 & 0 \\ 1.3060 & -2.1320 & 0.9807 & 0 & 0 \\ 0 & 1.5950 & -3.1490 & 1.5470 & 0 \\ 0 & 0.0355 & 2.6320 & -4.2570 & 1.8550 \\ 0 & 0.00227 & 0 & 0.1636 & -0.1625 \end{bmatrix}$$

$$B = \begin{bmatrix} 0 & 0.0632 & 0.0838 & 0.1004 & 0.0063 \\ 0 & 0 & -0.1396 & -0.2060 & -0.0128 \end{bmatrix}^T$$

Remaining matrices are chosen as

$$D^T D = \begin{bmatrix} 3 & 0 & 0.7 & 0.7 & 0.7 \\ 0 & 3 & 0.7 & 0.7 & 0.7 \\ 0.7 & 0.7 & 3 & 0 & 0 \\ 0.7 & 0.7 & 0 & 3 & 0 \\ 0.7 & 0.7 & 0 & 0 & 3 \end{bmatrix}$$

$$R = I_2, \quad F = I_5$$

The initial and final times are selected as $t_0 = 0$ and $T = 1$.

The system is partitioned into two subsystems with $n_1 = 2$, $n_2 = 3$, and ε = 0.6. The small parameter ε is built into the problem. It can be roughly estimated from the strongest coupled matrix - in this case matrix B - producing $|b_{31}|/|b_{32}|$ = 0.0838 / 0.1396 = 0.6. The simulation results for the differential Riccati equation are presented in Table 5.4. After performing 7 iterations, we have obtained the accuracy of 10^{-4}. Since $(0.6)^{16}$ = 2.8 x 10^{-4} the estimate of the coupling parameter ε is quite good.

iteration	t = 0.25	t = 0.5	t = 1
7=optimal	1.7479	2.6297	4.7559
6	1.7480	2.6300	4.7564
5	1.7464	2.6268	4.7504
4	1.7607	2.6548	4.8039
3	1.7740	2.6792	4.8423
2	1.7878	2.7484	5.1674
1	1.7150	2.4314	3.9356
0	1.5446	1.6680	0.74701

Table 5.4 Simulation result for the element $P_{11}(t)$ of the Riccati differential equation.

It can be seen, that the proposed algorithm converges very rapidly (only 7 iterations are required for the accuracy of 10^{-4} despite of the relatively big value of the coupling parameter ε).

Appendix 5.1

A Seventh Order Model of the Single-Machine-Infinite Bus System

In this appendix we analyze a seventh order model of the single-machine-infinite bus system, given in Figure 5.1. A five cycle 3 phase fault is applied on circuit "a" close to bus 2, and is cleared by opening circuit "a".

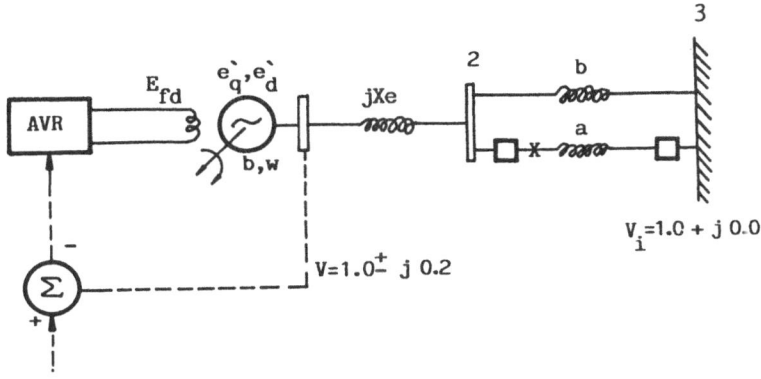

Figure 5.1 Single machine–infinite bus system

AVR is the standard IEEE type 1 voltage regulator (IEEE Committee Report, 1968) given in Figure 5.2. Table 5.5 contains the numerical values.

T_A = 0.06 s	K_E = -0.0445
T_E = 0.5 s	K_F = 0.16
T_F = 1.0 s	A_{SAT} = 0.001123
K_A = 25.0 s	B_{SAT} = 0.3043

Table 5.5 Voltage regulator constant

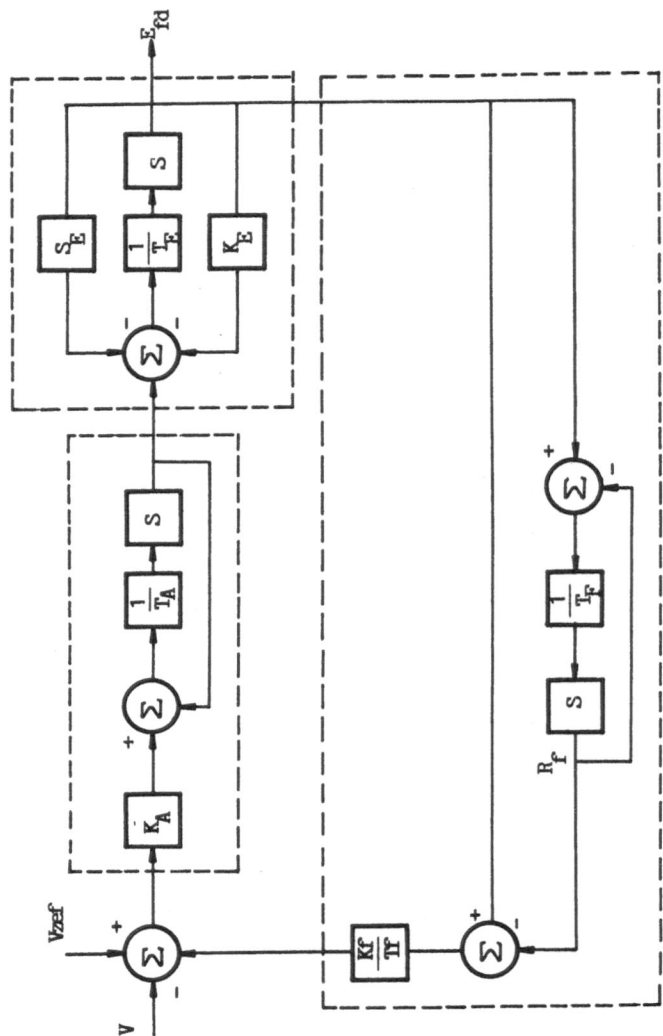

Figure 5.2 IEEE type 1 voltage regulator

In this type of voltage regulators commonly used state variables are separable. The system eigenvalues : -8.53 ± j8.22, -3.93, -0.326 ± j0.56, -0.86 ± j8.37 indicate that there should be two slow and five fast states. The slow variables, a $\Delta e'q$ (component of voltage behind transistent response due to direct axis flux linkages, with the field windings), ΔR_f (feedback compensator state) and the fast variable $\Delta \delta$ (machine angle) (see Kokotović, Allemong, Winkelman and Chow, 1980) are given in Figure 5.3-5.5, respectively.

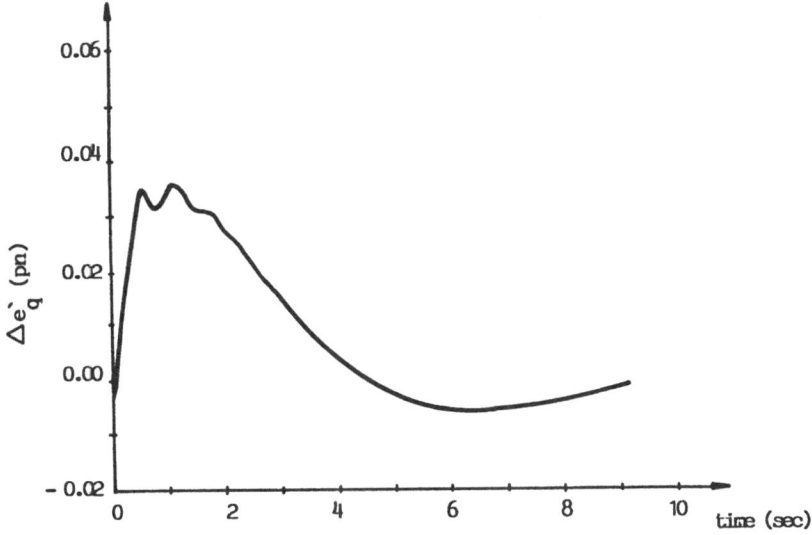

Figure 5.3 Slow variable $\Delta e'q$

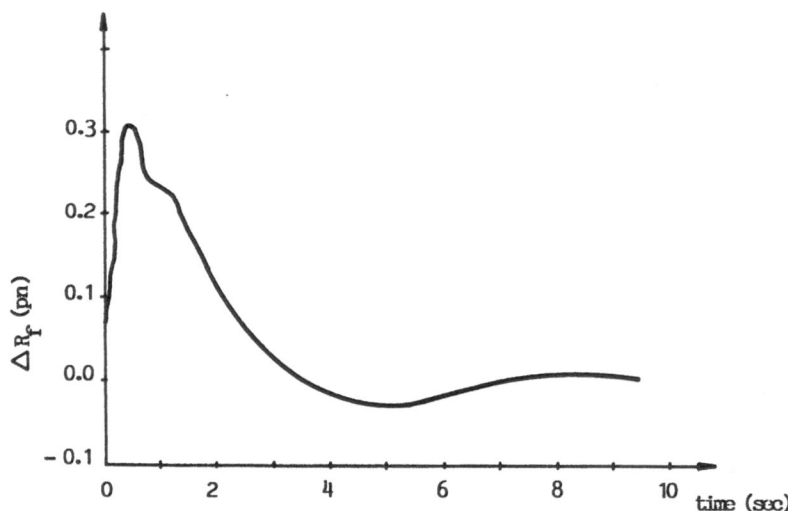

Figure 5.4 Slow variable ΔR_f

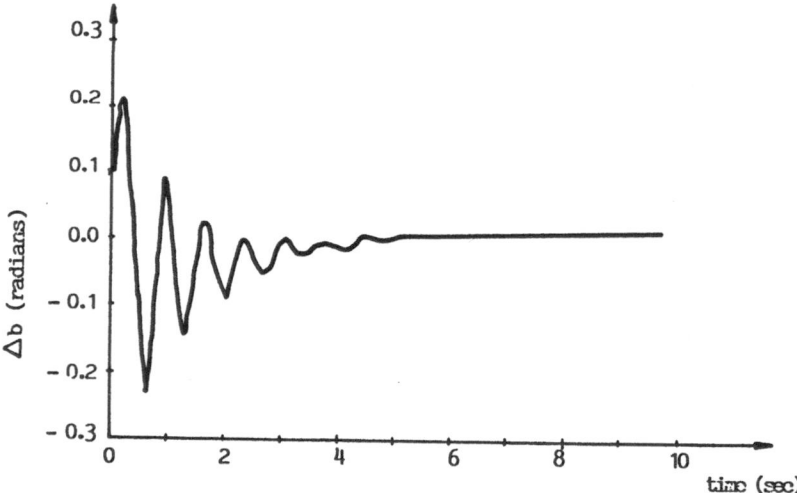

Figure 5.5 Fast variable $\Delta \delta$

APPLICATION TO THE DIFFERENTIAL GAMES

The recursive reduced order approach to the game theory of singularly perturbed and weakly coupled systems will be presented in the context of the Nash strategies for the weakly coupled systems only (Petrović and Gajić, 1988). Similar algorithm can be developed for the singularly perturbed linear-quadratic Nash games. The recursive approach to the other types of the differential games with small parameters (including high gain feedback and cheap control problems) is under consideration in (Lee, 1989).

6.1 Weakly Coupled Linear-Quadratic Nash Games

The linear quadratic Nash game strategies of large scale weakly interconnected (coupled) systems were studied in (Ozguner and Perkins, 1977) by means of a power series expansion method with respect to a small coupling parameter ε. This approach, originated in (Kokotović, Perkins, Cruz and D'Ans, 1969), is not recursive in its application and can be inferior compared to the hierarchical type decentralized control method (especially when ε is not very small), as was pointed out in (Mahmoud, 1978). In this section we develop a new recursive technique which will recover the importance of ideas presented in (Kokotović, Perkins, Cruz and D'Ans, 1969). Motivated by previous results for singularly perturbed systems, (Gajić, 1986), we have shown that weak coupling produces algebraic problems similar to those of (Gajić, 1986) and the fixed point method used in (Gajić, 1986) is very efficient in this case also.

As a matter of fact, we have developed an algorithm which converges very rapidly to the exact, nonnegative definite stabilizing solution of the coupled algebraic Riccati equations and thus to the optimal linear Nash strategies, even in the case when ε is not very small.

A controlled linear dynamic system under consideration is given by

$$\dot{x} = A(\varepsilon)x + B_1(\varepsilon)u_1 + B_2(\varepsilon)u_2 \tag{6.1}$$

where $x \in R^n$ is a state vector, $u_1 \in R^{m_1}$ and $u_2 \in R^{m_2}$ are control inputs, $A(\varepsilon)$, $B_1(\varepsilon)$ and $B_2(\varepsilon)$ are bounded matrix functions of a small parameter ε with compatible dimensions.

A quadratic type functional is associated with each control agent

$$J_1 = \int_0^\infty \left[x^T Q_1(\varepsilon)x + u_1^T R_1(\varepsilon)u_1 + u_2^T R_{12}(\varepsilon)u_2 \right] dt \tag{6.2a}$$

$$J_2 = \int_0^\infty \left[x^T Q_2(\varepsilon)x + u_1^T R_{21}(\varepsilon)u_1 + u_2^T R_2(\varepsilon)u_2 \right] dt \tag{6.2b}$$

where the weighting matrices are symmetric satisfying $Q_i(\varepsilon) \geq 0$, $R_i(\varepsilon) > 0$, $i = 1, 2$, $R_{ij}(\varepsilon) \geq 0$, $i = 1, 2$, $j = 1, 2$.

The optimal solution to the given problem with the conflict of interest and simultaneous decision making, (Starr and Ho, 1969), leads to so called Nash strategies u_1^* and u_2^* satisfying

$$J_1(u_1^*, u_2^*) \leq J_1(u_1, u_2^*) \tag{6.3a}$$

$$J_2(u_1^*, u_2^*) \leq J_2(u_1^*, u_2) \tag{6.3b}$$

It was shown in (Starr and Ho, 1969) that the optimal closed loop strategies are given by

$$u_i^* = R_i^{-1}(\varepsilon)B_i^T(\varepsilon)K_i(\varepsilon)x, \qquad i = 1, 2 \tag{6.4}$$

where K_i 's satisfy coupled algebraic Riccati equations

$$K_1(\varepsilon)A(\varepsilon) + A^T(\varepsilon)K_1(\varepsilon) + Q_1(\varepsilon) - K_1(\varepsilon)S_1(\varepsilon)K_1(\varepsilon) - K_1(\varepsilon)S_2(\varepsilon)K_2(\varepsilon)$$

$$- K_2(\varepsilon)S_2(\varepsilon)K_1(\varepsilon) + K_2(\varepsilon)Z_2(\varepsilon)K_2(\varepsilon) = 0 = N_1 \ (K_1, \ K_2) \tag{6.5a}$$

$$K_2(\varepsilon)A(\varepsilon) + A^T(\varepsilon) K_2(\varepsilon) + Q_2(\varepsilon) - K_2(\varepsilon)S_2(\varepsilon)K_2(\varepsilon) - K_2(\varepsilon)S_1(\varepsilon)K_1(\varepsilon)$$

$$- K_1(\varepsilon)S_1(\varepsilon)K_2(\varepsilon) + K_1(\varepsilon)Z_1(\varepsilon)K_1(\varepsilon) = 0 = N_2 \ (K_1, \ K_2) \tag{6.5b}$$

where

$$S_i(\varepsilon) = B_i(\varepsilon)R_i^{-1}(\varepsilon)B_i^T(\varepsilon), \qquad\qquad i = 1, \ 2$$

$$Z_i(\varepsilon) = B_i(\varepsilon)R_i^{-1}(\varepsilon)R_{ji}(\varepsilon)R_i^{-1}B_i^T(\varepsilon), \ i = 1, \ 2, \ j = 1, \ 2, \ i \neq j$$

The existence of the nonlinear optimal Nash strategies was established in (Basar, 1974), so that (6.4), in fact, are the best linear optimal strategies. Since a linear control law, from a practical point of view, is very desirable, the linear strategies (6.4) attract the attention of many researchers.

The existence of Nash strategies (6.4) and solutions of coupled Riccati equations (6.5) has been studied in (Papavassilopoulos, Medanić and Cruz, 1979), by means of Brower's fixed point theorem and by imposing norm conditions on the given matrices. In the recent paper (Li and Gajić, 1989), under control-oriented assumptions (Kucera, 1972, Wonham, 1968), the existence of nonnegative definite stabilizing solutions of (6.5) has been established.

It is important to point out that at the present time, there is no published method for finding stabilizing solutions of coupled algebraic Riccati equations (6.5). Some attempts in that direction have been made in (Bertrand, 1985, Papavassilopoulos and Olsder, 1984).

In this section, the Nash game problem is considered for a special case of weakly interconnected systems characterized by

$$A(\varepsilon) = \begin{bmatrix} A_1(\varepsilon) & \varepsilon A_{12}(\varepsilon) \\ \varepsilon A_{21}(\varepsilon) & A_2(\varepsilon) \end{bmatrix}$$

$$B_1(\varepsilon) = \begin{bmatrix} B_{11}(\varepsilon) \\ \varepsilon B_{21}(\varepsilon) \end{bmatrix}, \qquad B_2(\varepsilon) = \begin{bmatrix} \varepsilon B_{12}(\varepsilon) \\ B_{22}(\varepsilon) \end{bmatrix}$$

$$Q_1(\varepsilon) = \begin{bmatrix} U_1(\varepsilon) & \varepsilon U_{12}(\varepsilon) \\ \varepsilon U_{12}^T(\varepsilon) & \varepsilon^2 U_2(\varepsilon) \end{bmatrix}, \qquad Q_2(\varepsilon) = \begin{bmatrix} \varepsilon^2 V_1(\varepsilon) & \varepsilon V_{12}(\varepsilon) \\ \varepsilon V_{12}^T(\varepsilon) & V_2(\varepsilon) \end{bmatrix}$$

This partition decomposes the state vector x in two vectors $x_1 \in R^{n_1}$ and $x_2 \in R^{n_1}$ such that $n_1 + n_2 = n$. Since the small coupling parameter ε can not change the basic structures of the subsystems by destroying their main properties (otherwise we can not talk about the weak coupling), it is very natural to adopt the following form for the subsystem matrices.

Assumption 6.1. (Weak coupling assumption)

$$A_i(\varepsilon) = A_{io} + \varepsilon A_{oi}(\varepsilon)$$

$$B_{ii}(\varepsilon) = B_{io} + \varepsilon B_{oi}(\varepsilon)$$

$$U_i(\varepsilon) = U_{io} + \varepsilon U_{oi} \qquad\qquad i = 1, 2$$

$$V_2(\varepsilon) = V_{2o} + \varepsilon V_{o2}(\varepsilon)$$

$$R_i(\varepsilon) = R_{io} + \varepsilon R_{oi}(\varepsilon)$$

where $A_{oi}(\varepsilon)$, $B_{oi}(\varepsilon)$, $R_{oi}(\varepsilon)$, $i = 1, 2$, $U_{oi}(\varepsilon)$ and $V_{o2}(\varepsilon)$ are continuous functions of ε, whereas A_{io}, B_{io}, R_{io}, $i = 1, 2$ and U_{io}, V_{2o} are independent of ε.

In order to simplify the algebra, we will assume, without loss of generality, that $U_{12}(\varepsilon) = 0$, $V_{12}(\varepsilon) = 0$, $R_{12}(\varepsilon) = 0$, $R_{21}(\varepsilon) = 0$, $U_2(\varepsilon) = 0$, $V_1(\varepsilon) = 0$, $B_{12}(\varepsilon) = 0$, $B_{21}(\varepsilon) = 0$. Note that we are studying a more general case than the one studied in (Ozguner and Perkins, 1977) because of the ε-dependence of the problem matrices. In addition we do not need to impose the analyticity assumption with respect to ε, which must be done for the power series expansion method.

The following scaling of $K_1(\varepsilon)$ and $K_2(\varepsilon)$ is consistent with the nature of the solution of (6.5)

$$K_1(\varepsilon) = \begin{bmatrix} M_1(\varepsilon) & \varepsilon M_{12}(\varepsilon) \\ \varepsilon M_{12}^T(\varepsilon) & \varepsilon^2 M_2(\varepsilon) \end{bmatrix}, \; K_2(\varepsilon) = \begin{bmatrix} \varepsilon^2 N_1(\varepsilon) & \varepsilon N_{12}(\varepsilon) \\ \varepsilon N_{12}^T(\varepsilon) & N_2(\varepsilon) \end{bmatrix} \quad (6.6)$$

The very well known ε-decoupling method (Kokotović, Perkins, Cruz and D'Ans, 1969), based on the power series expansion with respect to ε, will convert the given full order problem (6.5) to a family of reduced order problems (Ozguner and Perkins, 1977). However, the power series expansion method is not recursive in nature and in the case when we are interested in high order of accuracy or when ε is not very small, the size of the required computations can be considerable. Moreover, when the problem matrices are functions of ε, the power series method demands the analyticity of all matrices. On the other hand, the expansion of quadratic terms (for example, $K_1(\varepsilon)B_1(\varepsilon)R_1^{-1}(\varepsilon)B_1^T(\varepsilon)K_1(\varepsilon)$) will produce an enormous number of terms, so that the reduced order advantage of the series expansion method becomes questionable. The presence of a small parameter ε will be exploited in the next section from a different point of view, leading to the recursive scheme for the solution of (6.5). Since the proposed method is of the fixed-point type, the boundness of all problem matrices over a compact set $\varepsilon \in [0, \varepsilon_1]$ has to be imposed. This is a much milder condition than the analyticity requirement of the power series expansion method.

6.2 Solution of Coupled Algebric Riccati Equations

Partitioning (6.5) compatibly with (6.6), we get the following set of equations

$$
M_1(\varepsilon)A_1(\varepsilon) + A_1^T(\varepsilon)M_1(\varepsilon) + U_1(\varepsilon) - M_1(\varepsilon)S_{11}(\varepsilon)M_1(\varepsilon) + \varepsilon^2\{M_{12}(\varepsilon)A_{21}(\varepsilon) +
$$
$$
A_{21}^T(\varepsilon)M_{12}^T(\varepsilon) - M_{12}(\varepsilon)S_{22}(\varepsilon)N_{12}^T(\varepsilon) - N_{12}(\varepsilon)S_{22}(\varepsilon)M_{12}^T(\varepsilon)\} = 0 \tag{6.7a}
$$

$$
M_1(\varepsilon)A_{12}(\varepsilon) + M_{12}(\varepsilon)A_2(\varepsilon) - M_1(\varepsilon)S_{11}(\varepsilon)M_{12}(\varepsilon) - M_{12}(\varepsilon)S_{22}(\varepsilon)N_2(\varepsilon) -
$$
$$
\varepsilon^2\{N_{12}(\varepsilon)S_{22}(\varepsilon)M_2(\varepsilon) - A_{21}^T(\varepsilon)M_2(\varepsilon)\} + A_1^T(\varepsilon)M_{12}(\varepsilon) = 0 \tag{6.7b}
$$

$$
M_2(\varepsilon)A_2(\varepsilon) + A_2^T(\varepsilon)M_2(\varepsilon) - M_2(\varepsilon)S_{22}(\varepsilon)N_2(\varepsilon) - N_2(\varepsilon)S_{22}(\varepsilon)M_2(\varepsilon)
$$
$$
+ M_{12}^T(\varepsilon)A_{12}(\varepsilon) + A_{12}^T(\varepsilon)M_{12}(\varepsilon) - M_{12}^T(\varepsilon)S_{11}(\varepsilon)M_{12}(\varepsilon) = 0 \tag{6.7c}
$$

$$
N_1(\varepsilon)A_1(\varepsilon) + A_1^T(\varepsilon)N_1(\varepsilon) - N_1(\varepsilon)S_{11}(\varepsilon)M_1(\varepsilon) - M_1(\varepsilon)S_{11}(\varepsilon)N_1(\varepsilon)
$$
$$
+ N_{12}(\varepsilon)A_{21}(\varepsilon) + A_{21}^T(\varepsilon)N_{12}^T(\varepsilon) - N_{12}(\varepsilon)S_{22}(\varepsilon)N_{12}^T(\varepsilon) = 0 \tag{6.7d}
$$

$$
\varepsilon^2 N_1(\varepsilon)A_{12}(\varepsilon) + N_{12}(\varepsilon)A_2(\varepsilon) - N_{12}(\varepsilon)S_{22}(\varepsilon)N_2(\varepsilon) - \varepsilon^2 N_1(\varepsilon)S_{11}(\varepsilon)M_{12}(\varepsilon)
$$
$$
- M_1(\varepsilon)S_{11}(\varepsilon)N_{12}(\varepsilon) + A_{21}^T(\varepsilon)N_2(\varepsilon) + A_1^T(\varepsilon)N_{12}(\varepsilon) = 0 \tag{6.7e}
$$

$$
N_2(\varepsilon)A_2(\varepsilon) + A_2^T(\varepsilon)N_2(\varepsilon) + V_2(\varepsilon) - N_2(\varepsilon)S_{22}(\varepsilon)N_2(\varepsilon) + \varepsilon^2\{N_{12}^T(\varepsilon)A_{12} +
$$
$$
A_{12}^T(\varepsilon)N_{12}(\varepsilon) - N_{12}^T(\varepsilon)S_{11}(\varepsilon)M_{12}(\varepsilon) - M_{12}^T(\varepsilon)S_{11}(\varepsilon)N_{12}(\varepsilon)\} = 0 \tag{6.7f}
$$

where

$$S_{ii}(\varepsilon) = B_{ii}(\varepsilon)R_i^{-1}(\varepsilon)B_{ii}^T(\varepsilon), \qquad\qquad i = 1, 2$$

6.2.1 Zeroth-Order Approximation

Let us define the $0\ (\varepsilon^2)$ perturbation of (6.7) as

$$\underline{M}_1(\varepsilon)A_1(\varepsilon) + A_1^T(\varepsilon)\underline{M}_1(\varepsilon) + U_1(\varepsilon) - \underline{M}_1(\varepsilon)S_{11}(\varepsilon)\underline{M}_1(\varepsilon) = 0 \qquad\qquad (6.8a)$$

$$\underline{M}_{12}(\varepsilon)D_2(\varepsilon) + D_1(\varepsilon)^T\underline{M}_{12}(\varepsilon) = -\underline{M}_1(\varepsilon)A_{12}(\varepsilon) \qquad\qquad (6.8b)$$

$$\underline{M}_2(\varepsilon)D_2(\varepsilon) + D_2(\varepsilon)^T\underline{M}_2(\varepsilon) = \underline{M}_{12}^T(\varepsilon)S_{11}(\varepsilon)\underline{M}_{12}(\varepsilon) - \underline{M}_{12}^T(\varepsilon)A_{12}(\varepsilon)$$

$$- A_{12}^T(\varepsilon)\underline{M}_{12}(\varepsilon) \qquad\qquad (6.8c)$$

$$\underline{N}_1(\varepsilon)D_1(\varepsilon) + D_1^T(\varepsilon)\underline{N}_1(\varepsilon) = \underline{N}_{12}(\varepsilon)S_{22}(\varepsilon)\underline{N}_{12}^T(\varepsilon) - \underline{N}_{12}(\varepsilon)A_{21}(\varepsilon)$$

$$- A_{21}^T(\varepsilon)M_{12}^T(\varepsilon) \qquad\qquad (6.8d)$$

$$\underline{N}_{12}(\varepsilon)D_2(\varepsilon) + D_1^T(\varepsilon)\underline{N}_{12}(\varepsilon) = - A_{21}^T(\varepsilon)\underline{N}_2(\varepsilon) \qquad\qquad (6.8e)$$

$$\underline{N}_2(\varepsilon)A_2(\varepsilon) + A_2^T(\varepsilon)\underline{N}_2(\varepsilon) + V_2(\varepsilon) - \underline{N}_2(\varepsilon)S_{22}(\varepsilon)\underline{N}_2(\varepsilon) = 0 \qquad\qquad (6.8f)$$

where

$$D_1(\varepsilon) = A_1(\varepsilon) - S_{11}(\varepsilon)\underline{M}_1(\varepsilon)$$

$$D_2(\varepsilon) = A_2(\varepsilon) - S_{22}(\varepsilon)\underline{N}_2(\varepsilon)$$

This system of equations has decoupled form and can be solved like two lower-order Riccati equations (6.8a), (6.8f) and four low-order

Lyapunov equations (6.8b)-(6.8e). The nonnegative definite stabilizing solution of (6.8a) and (6.8f) exist under the well-known stabilizability-detectability assumption (Kucera, 1972, Wonham, 1968).

Assumption 6.2. The triples $(A_1(0), B_1(0), \sqrt{U_1}(0))$ and $(A_2(0), B_2(0), \sqrt{V_2}(0))$ are stabilizable-detectable.

Under the same assumption, the unique solution of (6.6b)-(6.8e) exist since $D_1(\varepsilon)$ and $D_2(\varepsilon)$ are stable matrices, (Kucera, 1972, Wonham, 1968).

6.2.2 Solution of Higher-Order Accuracy

The zeroth-order solutions $\underline{M}(\varepsilon)$ and $\underline{N}(\varepsilon)$ are $O(\varepsilon^2)$ close to the exact ones. Then the exact solutions can be sought in the form

$$K_1(\varepsilon) = \begin{bmatrix} M_1(\varepsilon) + \varepsilon^2 E_1(\varepsilon) & \varepsilon\{M_{12}(\varepsilon) + \varepsilon^2 E_{12}(\varepsilon)\} \\ \varepsilon\{\underline{M}_{12}(\varepsilon) + \varepsilon^2 E_{12}(\varepsilon)\}^T & \varepsilon^2\{\underline{M}_2(\varepsilon) + \varepsilon^2 E_2(\varepsilon)\} \end{bmatrix} \tag{6.9a}$$

$$K_2(\varepsilon) = \begin{bmatrix} \varepsilon^2\{\underline{N}_1(\varepsilon) + \varepsilon^2 G_1(\varepsilon)\} & \varepsilon\{\underline{N}_{12}(\varepsilon) + \varepsilon^2 G_{12}(\varepsilon)\} \\ \varepsilon\{\underline{N}_{12}(\varepsilon) + \varepsilon^2 G_{12}(\varepsilon)\}^T & \underline{N}_2(\varepsilon) + \varepsilon^2 G_2(\varepsilon) \end{bmatrix} \tag{6.9b}$$

Obviously, $O(\varepsilon^k)$ approximations of $E(\varepsilon)$'s and $G(\varepsilon)$'s will produce $O(\varepsilon^{k+2})$ approximations of required solutions, which is why we are interested in finding convenient form for these error terms and the appropriate algorithm for their solution.

Subtracting equations (6.8) from corresponding equations (6.7) and after doing some algebra we get the following expressions for the error equations:

$$E_1 D_1 + D_1^T E_1 = C_1 + \varepsilon^2 F_1 (E_1, E_{12}, G_{12}) \tag{6.10a}$$

$$E_1 D_{12} + E_{12} D_2 + D_1^T E_{12} - \underline{M}_{12} S_{22} G_2 = C_2 + \varepsilon^2 F_2 (E_1, E_{12}, E_2, G_{12}, G_2) \tag{6.10b}$$

$$E_{12}^T D_{12} + D_{12}^T E_{12} + E_2 D_2 + D_2^T E_2 - G_2 S_{22} \underline{M}_2 - \underline{M}_2 S_{22} G_2 = \varepsilon^2 F_3(E_{12}, E_2, G_2) \tag{6.10c}$$

$$G_1 D_1 + D_1^T G_1 + G_{12} D_{21} + D_{21}^T G_{12}^T - E_1 S_{11} \underline{N}_1 - \underline{N}_1 S_{11} E_1 = \varepsilon^2 F_4(E_1, G_{12}, G_2) \tag{6.10d}$$

$$G_{12} D_2 + D_1^T G_{12} + D_{21}^T G_2 - E_1 S_{11} \underline{N}_{12} = C_5 + \varepsilon^2 F_5(E_1, E_{12}, G_1, G_{12}, G_2) \tag{6.10e}$$

$$G_2 D_2 + D_2^T G_2 = C_6 + \varepsilon^2 F_6(E_{12}, G_{12}, G_2) \tag{6.10f}$$

where

$$D_{12} = D_{12}(\varepsilon) = A_{12}(\varepsilon) - S_{11}(\varepsilon) \underline{M}_{12}(\varepsilon)$$

$$D_{21} = D_{21}(\varepsilon) = A_{21}(\varepsilon) - S_{22}(\varepsilon) \underline{N}_{12}(\varepsilon)$$

Matrices F_i, $i = 1, 2, \ldots 6$ and constant matrices C_j are given in Appendix 6.1. In order to simplify notation the ε-dependence of the problem matrices in the equation (6.10) and in the remaining part of the chapter is omitted.

The weakly coupled and hierarchical structure of (6.10) can be exploited by proposing the following recursive scheme, which leads, after some algebra, to the six low-order completely decoupled Lyapunov equations

$$E_1^{(i+1)} D_1 + D_1^T E_1^{(i+1)} = \varepsilon^2 E_1^{(i)} S_{11} E_1^{(i)} - M_{12}^{(i)} D_{21}^{(i)} - D_{21}^{(i)^T} M_{12}^{(i)^T} \tag{6.11a}$$

$$G_2^{(i+1)} D_2 + D_2^T G_2^{(i+1)} = \varepsilon^2 G_2^{(i)} S_{22} G_2^{(i)} - N_{12}^{(i)^T} D_{12}^{(i)} - D_{12}^{(i)^T} N_{12}^{(i)} \tag{6.11f}$$

$$E_{12}^{(i+1)} D_2 + D_1^T E_{12}^{(i+1)} = -E_1^{(i+1)} D_{12}^{(i)} + M_{12}^{(i)} S_{22} G_2^{(i+1)} - D_{21}^{(i)^T} M_2^{(i)} \tag{6.11b}$$

$$G_{12}^{(i+1)} D_2 + D_1^T G_{12}^{(i+1)} = -D_{21}^{(i)^T} G_2^{(i+1)} + E_1^{(i+1)} S_{11} N_{12}^{(i)} - N_1^{(i)} D_{12}^{(i)} \tag{6.11e}$$

$$E_2^{(i+1)}D_2 + D_2^T E_2^{(i+1)} = M_2^{(i)}S_{22}G_2^{(i+1)} + G_2^{(i+1)}S_{22}M_2^{(i)} - E_{12}^{(i+1)T}D_{12}$$

$$\text{(6.11c)}$$

$$- D_{12}^T E_{12}^{(i+1)} + \varepsilon^2 E_{12}^{(i+1)T}S_{11}E_{12}^{(i+1)}$$

$$G_1^{(i+1)}D_1 + D_1^T G_1^{(i+1)} = E_1^{(i+1)}S_{11}N_1^{(i)} + N_1^{(i)}S_{11}E_1^{(i+1)} - G_{12}^{(i+1)}D_{21}$$

$$\text{(6.11d)}$$

$$- D_{21}^T G_{12}^{(i+1)T} + \varepsilon^2 G_{12}^{(i+1)}S_{22}G_{12}^{(i+1)T}$$

$$i = 0, 1, 2, 3, \dots\dots$$

with initial conditions chosen as

$$E_1^{(0)} = E_{12}^{(0)} = E_2^{(0)} = G_1^{(0)} = G_{12}^{(0)} = G_2^{(0)} = 0$$

where

$$M_{12}^{(i)} = \underline{M}_{12} + \varepsilon^2 E_{12}^{(i)}$$

$$N_{12}^{(i)} = \underline{N}_{12} + \varepsilon^2 G_{12}^{(i)}$$

$$N_1^{(i)} = \underline{N}_1 + \varepsilon^2 G_1^{(i)}$$

$$\qquad\qquad i = 1, 2, 3, \dots$$

$$M_2^{(i)} = \underline{M}_2 + \varepsilon^2 E_2^{(i)}$$

$$D_{12}^{(i)} = A_{12} - S_{11}M_{12}^{(i)}$$

$$D_{21}^{(i)} = A_{21} - S_{22}N_{12}^{(i)T}$$

These Lyapunov equations have to be solved in the given order, that is, first E_1 and G_2, then E_{12} and G_{12}, and finally E_2 and G_1.

The following theorem indicates the features of the proposed recursive scheme.

Theorem 6.1 Under imposed weak coupling and stabilizability and detectability assumptions, given algorithm (6.11) converges to the exact solution of the error terms, and thus of $K_1(\varepsilon)$ and $K_2(\varepsilon)$, with the rate of convergence of $O(\varepsilon^2)$, that is

$$\| E_j(\varepsilon) - E_j^{(i)}(\varepsilon) \| = o(\varepsilon^{2i})$$

$$\| G_j(\varepsilon) - G_j^{(i)}(\varepsilon) \| = O(\varepsilon^{2i}) \qquad j = 1, 2$$

$$\| E_{12}(\varepsilon) - E_{12}^{(i)}(\varepsilon) \| = O(\varepsilon^{2i}) \qquad i = 1, 2, 3, ...,$$

$$\| G_{12}(\varepsilon) - G_{12}^{(i)}(\varepsilon) \| = O(\varepsilon^{2i})$$

(6.12)

and

$$\| K_j(\varepsilon) - K_j^{(i)}(\varepsilon) \| = O(\varepsilon^{2i+2}) \qquad j = 1, 2, i = 0, 1, 2, .$$

Proof: As a starting point, we need to show the existence of a bounded solution of (6.10) in the neighborhood of $\varepsilon = 0$. By the implicit function theorem it is enough to show that the corresponding Jacobian is nonsingular at $\varepsilon = 0$. The Jacobian is given by

$$J(\varepsilon)\Big|_{\varepsilon=0} = \begin{bmatrix} \Gamma_1 & 0 & 0 & 0 & 0 & 0 \\ * & \Gamma_2 & 0 & 0 & 0 & * \\ 0 & * & \Gamma_3 & 0 & 0 & * \\ * & 0 & 0 & \Gamma_1 & * & 0 \\ * & 0 & 0 & 0 & \Gamma_2 & * \\ 0 & 0 & 0 & 0 & 0 & \Gamma_3 \end{bmatrix} \qquad (6.13)$$

where the asterisk denotes terms which are not important for a nonsingularity of the Jacobian. Γ's are given by the Kronecker product representation

$$\Gamma_i = I_{n_i} \times D_i^T(0) + D_i^T(0) \times I_{n_i} \qquad i = 1, 3$$

$$\Gamma_2 = I_{n_2} \times D_2^T(0) + D_1^T(0) \times I_{n_1}$$

where I_{n_1} and I_{n_2} are identity matrices. Under Assumptions 6.1 and 6.2, $D_1(0)$ and $D_2(0)$ are stable matrices for any sufficiently small $\varepsilon \in \left[0, \varepsilon_2\right]$ and by well known properties of the Kronecker product (Lanscaster and Tismenetsky, 1985), so are matrices Γ_1, Γ_2, and Γ_3. It is easy to see that the nonsingularity of the Jacobian is guaranteed by the nonsingularity of Γ_1, Γ_2 and Γ_3.

The second step in the proof of the given theorem is to give an estimate of the rate of convergence.

For i = 0, (6.10a) and (6.11a) imply

$$(E_1 - E_1^{(1)})D_1 + D_1^{T}(E_1 - E_1^{(1)}) = \varepsilon^2 F_1(E_1, E_{12}, G_{12})$$

which by stability of D_1 and the existence of the bounded solution of (6.10) gives

$$\| E_1 - E_1^{(1)} \| = O(\varepsilon^2) \tag{6.14a}$$

By the same arguments from (6.10f) and (6.11f) we have

$$\| G_2 - G_2^{(1)} \| = O(\varepsilon^2) \tag{6.14f}$$

Subtracting (6.11b) from (6.10b) and using (6.14a) and (6.14f) and the expression for F_3 (from Appendix 6.1) lead to

$$(E_{12} - E_{12}^{(1)}) D_2 + D_1^{T}(E_{12} - E_{12}^{(1)}) = O(\varepsilon^2)$$

which implies

$$\| E_{12} - E_{12}^{(1)} \| = O(\varepsilon^2) \tag{6.14b}$$

By the analogy (equations (6.10b) and (6.10e) have the similar form), (6.10e) and (6.11e) will produce

$$\| G_{12} - G_{12}^{(1)} \| = O(\varepsilon^2) \tag{6.14e}$$

Also, from (6.10c), (6.11c), (6.14a,b,e,f) and the Appendix 6.1 we have

$$(E_2 - E_2^{(1)})D_2 + D_2^{T}(E_2 - E_2^{(1)}) = O(\varepsilon^2)$$

that is

$$\| E_2 - E_2^{\cdot\,(1)} \| = O(\varepsilon^2) \tag{6.14c}$$

and by analogy, from (6.10d) and (6.11d) we get

$$\| G_1 - G_1^{(1)} \| = O(\varepsilon^2) \tag{6.14d}$$

Using these starting observations and forms of F_j's and C_j's it can be shown that

$$\| F_j - F_j^{(1)} \| = O(\varepsilon^{2i}), \quad j = 1, 2, \ i = 1, 2, 3, \ \ldots\ldots \tag{6.15}$$

For example, for $j = 1$

$$F_1 - F_1^{(1)} = (E_1 - E_1^{(1)})S_{11}E_1^{(1)} + E_1 S_{11}(E_1 - E_1^{(1)})$$

$$- (E_{12} - E_{12}^{(1)})D_{21} - D_{21}^T(E_{12} - E_{12}^{(1)})^T$$

$$+ (G_{12} - G_{12}^{(1)})S_{22}M_{12}^{(1)^T} + M_{12}S_{22}(G_{12} - G_{12}^{(1)})^T$$

so that for $i = 1$, from (6.14) we have $F_1 - F_1^{(1)} = O(\varepsilon^2)$, that is

$$(E_1 - E_1^{(2)})D_1 + D_1^T(E_1 - E_1^{(2)}) = \varepsilon^2(F_1 - F_1^{(1)}) = O(\varepsilon^4)$$

which implies that

$$(E_1 - E_1^{(2)}) = O(\varepsilon^4)$$

Continuing the same procedure we can verify (6.15), which by the existence of the bounded solutions of E's and G's will imply (6.12). Note that the solution of (6.11) exist at each iteration since the corresponding Jacobian is always given by (6.13), and thus nonsingular at $\varepsilon = 0$ in every iteration.

We would like to point out that the imposed form of solution (6.9) is an additional limiting factor for a small parameter ε. It was shown in (Li and Gajić, 1989), that under Assumption 6.2, the solution of (6.5) is nonnegative definite. Since the solution of (6.10) is only symmetric, (which can be easily seen from the form of corresponding equations) the small parameter ε has to be constrained to the set $\varepsilon \in \left[0, \varepsilon_3\right]$ such that $\forall \varepsilon$, $K_1(\varepsilon)$ and $K_2(\varepsilon)$ preserve required nonnegative definiteness. Thus, the presented method is applicable for $\varepsilon \in \left[0, \varepsilon^*\right]$, where $\varepsilon^* = \min\{\varepsilon_1, \varepsilon_2, \varepsilon_3, ...\}$. However, the limiting condition

$$\varepsilon^* = \min \{\varepsilon_1, \varepsilon_2, \varepsilon_3, ...\}$$

is present in the entire theory of small parameters (weak coupling and singular perturbations), it is both method-dependent and problem-dependent, and it is not a direct consequence of the procedure studied in this book.

Let us compare the proposed algorithm (6.11), based on the fixed point iteration, for weakly coupled systems, and the power series expansion algorithm for the same type of systems. The comparison is done for the case when the problem matrices are not functions of ε (which is in the favor of the power series expansion algorithm). The equations corresponding to (6.11) are given by, (Ozguner and Perkins, 1977)

$$M_1^{(i+1)}D_1 + D_1^T M_1^{(i+1)} = Z_1^{(0,1,2,...,i)} \tag{6.16a}$$

$$N_2^{(i+1)}D_2 + D_2^T N_2^{(i+1)} = Z_6^{(0,1,2,...,i)} \tag{6.16f}$$

$$M_{12}^{(i+1)}D_2 + D_1^T M_{12}^{(i+1)} = Z_2^{(0,1,2,...,i)} \tag{6.16b}$$

$$N_{12}^{(i+1)}D_2 + D_1^T G_{12}^{(i+1)} = Z_5^{(0,1,2,...,i)} \tag{6.16e}$$

$$M_2^{(i+1)}D_2 + D_2^T M_2^{(i+1)} = Z_3^{(0,1,2,...,i)} \tag{6.16c}$$

$$N_1^{(i+1)}D_1 + D_1^T N_1^{(i+1)} = Z_4^{(0,1,2,...,i)} \tag{6.16d}$$

where Z_j, $j = 1, 2,...6$ depend on the all previously obtained terms. For example

$$Z_1^{(0,1,2, ...,i)} = -(i+1)\{M_{12}^{(i)}A_{21} + A_{21}^T M_{12}^{(i)T}\} + \sum_{\substack{k=2 \\ k \text{ even}}}^{i-1}\binom{i+1}{k}M_1^{(i+1-k)}S_{11}M_1^{(k)}$$

$$+ \sum_{\substack{k=1 \\ k \text{ odd}}}^{i}\{\binom{i+1}{k}M_{12}^{(i+1-k)}S_{22}N_{12}^{(k)T} + N_{12}^{(i+1-k)}S_{22}M_{12}^{(k)}\} \tag{6.17}$$

Both approaches produce the same type of equations (Lyapunov ones), but in order to form the right hand side, for example of (6.11a), we have to perform only 3 matrix multiplications for every i, whereas for corresponding equation of the power series expansion the number of required matrix multiplications grows very quickly as i increases (6.17). Thus, the obvious advantages of the fixed point iteration approach are:

1) The size of required computation is considerably less, and since it does not grow per iteration, the proposed method is extremally efficient for obtaining the exact solution or the solution of very high accuracy.

2) The fixed point method is recursive in nature (the power series expansion method is not), and thus much easier to implement.

The approximations of the suboptimal Nash strategies (6.4) can be defined by

$$u_j^{(i)}(t) = - R_j^{-1}(\varepsilon)B_j^T(\varepsilon)K_j^{(i)}(\varepsilon)x(t), \quad j = 1, 2, \; i = 0, 1, 2, 3, \; .... \tag{6.18}$$

where

$$K_1^{(i)} = \begin{bmatrix} \underline{M}_1(\varepsilon) + \varepsilon^2 E_1^{(i)}(\varepsilon) & \varepsilon\{\underline{M}_{12}(\varepsilon) + \varepsilon^2 E_{12}(i)(\varepsilon)\} \\ \varepsilon\{\underline{M}_{12}(\varepsilon) + \varepsilon^2 E_{12}^{(i)}(\varepsilon)\}^T & \varepsilon^2\{\underline{M}_2(\varepsilon) + \varepsilon^2 E_2(i)(\varepsilon)\} \end{bmatrix} \tag{6.19a}$$

$$K_2^{(i)} = \begin{bmatrix} \varepsilon^2\{\underline{N}_1(\varepsilon) + \varepsilon^2 G_1^{(i)}(\varepsilon)\} & \varepsilon\{\underline{N}_{12}(\varepsilon) + \varepsilon^2 G_{12}^{(i)}(\varepsilon)\} \\ \varepsilon\{\underline{N}_{12}(\varepsilon) + \varepsilon^2 G_{12}^{(i)}(\varepsilon)\} & \underline{N}_2(\varepsilon) + \varepsilon^2 G_2^{(i)}(\varepsilon) \end{bmatrix} \tag{6.19b}$$

Then, by following the arguments of (Cruz and Chen, 1971), the cost approximations produce

$$J_j^{(i)}(u_1^{(i)}, u_2^{(i)}) = J_j(u_1^*, u_2^*) + O(\varepsilon^{2i+2}), \quad j = 1, 2, \; i = 0, 1, 2,... \tag{6.20}$$

The approximate cost functions for the other cases, when the control agents use the approximative strategies of the different order of accuracy (for example $u_1^{(p)}$ and $u_2^{(q)}$, $p \neq q$) can be obtained by using results of

(Cruz and Chen, 1971) also. But, since the proposed method is recursive in its nature, and thus very easy to implement, and since the amount of required computations is constant per iteration (does not grow with i) the accuracy of very high order can be achieved at a very low cost, so that the proposed method can be efficient for finding the exact solution as well.

Since the proposed algorithm defines the error of approximation similarly to the power series expansion, it can be easily seen, that the approximate Nash strategies (6.18) are also well posed in the sense of Khalil (1980).

6.3 Numerical Example

In order to demonstrate the efficiency of the proposed algorithm, we have run a fourth order example. Matrices A_1, A_{12}, A_{21}, A_2, B_{11} and B_{22} have been chosen randomly (standard deviation = 1, mean value = 0) and the matrices $R_1 = R_2 = U_1 = V_2 = I$ are chosen such that the required stabilizability-detectability assumptions are satisfied.

$$A_1 = \begin{bmatrix} -1.035 & -0.192 \\ 1.684 & -0.421 \end{bmatrix} \qquad A_{12} = \begin{bmatrix} -1.084 & 0.579 \\ 1.327 & -0.841 \end{bmatrix}$$

$$A_{21} = \begin{bmatrix} -1.370 & -0.533 \\ 1.069 & 0.835 \end{bmatrix} \qquad A_2 = \begin{bmatrix} -1.510 & -0.139 \\ 0.410 & 1.238 \end{bmatrix}$$

$$B_{11} = \begin{bmatrix} -1.019 & 0.602 \\ -0.912 & 1.329 \end{bmatrix} \qquad B_{22} = \begin{bmatrix} -1.641 & 0.330 \\ 1.068 & 0.243 \end{bmatrix}$$

$$U_1 = V_2 = R_1 = R_2 = \begin{bmatrix} 1 & 0 \\ 0 & 1 \end{bmatrix}$$

The simulation results for different values of a coupling parameter ε are given in Table 6.1. Since we do not know the exact solution of the equation (6.5), (no method available in the literature at the present time), the error is defined as

$$e^{(i)} = \max \left\{ \; \| \; N_1(K_1^{(i)}, \; K_2^{(i)}) \; \|_\infty, \; \| \; N_2(K_1^{(i)}, \; K_2^{(i)}) \; \|_\infty \right\}$$

In the second table we have shown the propagation of the error per iteration when $\varepsilon = 0.1$.

ε	i = number of required iterations such that $e^{(i)} < 10^{-10}$
0.8	16
0.6	11
0.4	8
0.2	5
0.1	4
0.05	3
0.01	2
0.001	1

Table 6.1. Dependence of number of iterations on ε

$\varepsilon = 0.1$	error $e^{(i)}$
i	
0	0.89662×10^{-2}
1	0.65481×10^{-4}
2	0.10349×10^{-6}
3	0.40663×10^{-9}
4	0.92572×10^{-11}

Table 6.2 Propagation of the error per iteration for a constant value of ε

The results from Table 6.1 strongly support the necessity of the existence of the recursive scheme for the solution of weakly coupled linear-quadratic Nash game problem, since unless ε is very small, the zeroth and first order approximations are far from the optimal solution.

Results from Table 6.2 verify, for this particular example, the conclusions of Theorem 6.1, that is, the rate of convergence of the proposed algorithm is $O(\varepsilon^2) = O(10^{-2})$.

Therefore, the solution to the Nash strategies of weakly interconnected systems can be obtained up to an arbitrary accuracy by performing iterations on the Lyapunov equations corresponding to the local subsystem problems. This idea hopefully can be extended to the generalized weak coupling problem (Sezar and Šiljak, 1986) and to the weakly coupled nonlinear systems (Kokotović and Singh, 1971).

Appendix 6.1

$$F_1 = E_1 S_{11} E_1 + \underline{M}_{12} S_{22} G_{12}{}^T + G_{12} S_{22} \underline{M}_{12}{}^T - E_{12} D_{21} - D_{21}{}^T E_{12}{}^T +$$

$$\varepsilon^2 (E_{12} S_{22} G_{12}{}^T + G_{12} S_{22} E_{12}{}^T)$$

$$F_2 = E_{12} S_{22} G_2 + E_1 S_{11} E_{12} + G_{12} S_{22} \underline{M}_2 - D_{21}{}^T E_2 + \varepsilon^2 G_{12} S_{22} E_2$$

$$F_3 = E_{12} S_{11} E_{12} + E_2 S_{22} G_2 + G_2 S_{22} E_2$$

$$F_4 = G_{12} S_{22} G_{12}{}^T + E_1 S_{11} G_1 + G_1 S_{11} E_1$$

$$F_5 = E_1 S_{11} G_{12} + G_{12} S_{22} G_2 + \underline{N}_1 S_{11} G_{12} - G_1 D_{12} + \varepsilon^2 G_1 S_{11} E_{12}$$

$$F_6 = G_2 S_{22} G_2 + E_{12}{}^T S_{11} \underline{N}_{12} + \underline{N}_{12}{}^T S_{11} E_{12} - G_{12}{}^T D_{12} - D_{12}{}^T G_{12}$$

$$+ \varepsilon^2 (E_{12}{}^T S_{11} E_{12} + G_{12}{}^T S_{11} E_{12})$$

$$C_1 = -\underline{M}_{12} A_{12} - A_{21}{}^T \underline{M}_{12}{}^T + \underline{M}_{12} S_{22} \underline{N}_{12}{}^T + \underline{N}_{12} S_{22} \underline{M}_{12}{}^T$$

$$C_2 = - D_{21}{}^T \underline{M}_2$$

$$C_5 = - \underline{N}_1 D_{12}$$

$$C_6 = - \underline{N}_{12}{}^T A_{12} - A_{12}{}^T \underline{N}_{12} + \underline{M}_{12}{}^T S_{11} \underline{N}_{12} + \underline{N}_{12}{}^T S_{11} \underline{M}_{12}$$

LINEAR DISCRETE WEAKLY COUPLED CONTROL SYSTEMS

7.1 Optimal Reduced-Order Recursive Solution of the Weakly Coupled Discrete Riccati Equation

The linear weakly coupled discrete systems have not been studied in the literature yet. This is due to the fact that the partitioned form of the main equation of the optimal linear control theory - Riccati equation, has a very complicated form in the discrete time domain. In this section that problem is overcome by the use of the bilinear transformation, which is applicable under quite mild assumption, so that the solution of the discrete algebraic Riccati equation of weakly coupled systems is obtained by using known results for the corresponding continuous-time equation.

The algebraic Riccati equation of weakly coupled linear discrete systems is given by

$$P = A^T PA + Q - A^T PB(B^T PB + R)^{-1} B^T PA, \quad R > 0, \ Q \geq 0 \qquad (7.1)$$

where

$$A = \begin{bmatrix} A_1 & \varepsilon A_2 \\ \varepsilon A_3 & A_4 \end{bmatrix}, \ B = \begin{bmatrix} B_1 & \varepsilon B_2 \\ \varepsilon B_3 & B_4 \end{bmatrix}, \ Q = \begin{bmatrix} Q_1 & \varepsilon Q_2 \\ \varepsilon Q_2^T & Q_3 \end{bmatrix}, \ R = \begin{bmatrix} R_1 & 0 \\ 0 & R_2 \end{bmatrix}$$

and ε is a small coupling parameter. Due to block dominant structure of the problem matrices the required solution P has the form

$$
P = \begin{bmatrix} P_1 & \varepsilon P_2 \\ \varepsilon P_2^T & P_3 \end{bmatrix} \tag{7.2}
$$

The main goal in the theory of the weakly coupled systems is to obtain the required solution in terms of the reduced-order problems, namely subsystems. In the case of the algebraic weakly coupled discrete Riccati equation, the inversion of the partitioned matrix $(B^TPB + R)$ will produce a lot of terms and make corresponding approach computationally very involved even though one is faced with the reduced-order numerical problems.

In order to solve this problem, we have used the bilinear transformation introduced in (Kondo and Furuta, 1986) to transform the discrete Riccati equation (7.1) into a continuous-time algebraic Riccati equation of the form

$$
A_c^T P_c + P_c A_c + Q_c - P_c S_c P_c = 0, \qquad S_c = B_c R_c^{-1} B_c^T \tag{7.3}
$$

such that the solution of (7.1) is equal to the solution of (7.3). It is shown that the equation (7.3) preserves the structure of weakly coupled systems. It can be solved in terms of the reduced-order problems very efficiently by using the fixed point type method developed in Chapter 2 which converges to the required solution with the rate of convergence of $O(\varepsilon^2)$.

7.1.1 Computational Algorithm

Since the proposed algorithm for the discrete algebraic Riccati equation combines features of the bilinear transformation and fixed point algorithm developed in Chapter 2 for weakly coupled continuous algebraic Riccati equation, we will briefly summarize main results from (Kondo and Furuta, 1986). The bilinear transformation states that equations (7.1) and (7.3) have the same solution if the following hold

$$A_c = I - 2D^{-T} \tag{7.4a}$$

$$S_c = 2(I + A)^{-1}S_d D^{-1}, \qquad S_d = BR^{-1}B^T \tag{7.4b}$$

$$Q_c = 2D^{-1}Q(I + A)^{-1} \tag{7.4c}$$

$$D = (I + A^T) + Q(I + A)^{-1}S_d \tag{7.4d}$$

assuming that $(I + A)^{-1}$ exists. It is shown in (Bar-Ness and Halbersberg, 1980) that the matrix D is invertible. The physical interpretation of the transformation between the continuous type and discrete type algebraic Riccati equation is discussed in (Kondo and Furuta, 1986). Thus, the proposed algorithm will be valid under the assumption that the system matrix A has no eigenvalues located at -1. It is important to point out that the eigenvalues located in the neighbourhood of -1 will produce ill-conditioning with respect to the matrix inversion and make the algorithm numerically unstable. In other words, the proposed algorithm is efficient when the eigenvalues of the system matrix A are far enough from the point -1.

It can be easily verified that the weakly coupled structure of matrices defined in (7.1) will produce the weakly coupled structure of transformed matrices given in (7.4). That can be concluded from the fact that

$$(I + A)^{-1} = \begin{bmatrix} O(1) & O(\varepsilon) \\ O(\varepsilon) & O(1) \end{bmatrix}$$

Since S_d from (7.4b) and Q have the weakly coupled structure so does D from (7.4d). The inverse of D, that is D^{-1}, is also in the weakly coupled form. Then from (7.4a) and (7.4c) the weakly coupled structure of matrices A_c and Q_c follows directly since they are given in terms of sums and/or poducts of matrices having weakly coupled structure. Let us introduce compatible partitions of corresponding matrices

$$A_c = \begin{bmatrix} A_{11} & \varepsilon A_{12} \\ \varepsilon A_{21} & A_{22} \end{bmatrix}, \qquad\qquad S_c = \begin{bmatrix} S_{11} & \varepsilon S_{12} \\ \varepsilon S_{12}^T & S_{22} \end{bmatrix}$$

$$Q_c = \begin{bmatrix} Q_{11} & \varepsilon Q_{12} \\ \varepsilon Q_{12}^T & Q_{22} \end{bmatrix}, \qquad\qquad P_c = P = \begin{bmatrix} P_1 & \varepsilon P_2 \\ \varepsilon P_2^T & P_3 \end{bmatrix}$$

Note that these partitions have to be performed by a computer only in the process of calculations and there is no need for the corresponding analytical expressions.

Solution of (7.3) can be found in terms of the reduced-order problems by imposing standard stabilizability-detectability assumptions on subsystems. The efficient fixed point reduced-order algorithm for solving (7.3) is obtained in Section 2.3.2.

Unique positive semi-definite stabilizing solution of (7.3) exists under the following assumption (see Chapter 2).

Assumption 7.1. Triples $\left(A_{ii}, \sqrt{S_{ii}}, \sqrt{Q_{ii}}\right)$, $i = 1, 2$ are stabilizable and detectable.

Thus, the proposed algorithm for the reduced-order solution of the discrete algebraic Riccati equation has the following form:

1) Transform (7.1) into (7.3) by using (7.4).

2) Solve (7.3) by using the reduced-order algorithm from Section 2.3.2.

7.1.2. Case Study: Discrete Model of the Distillation Column

A real world physical example (distillation column studied by Kautsky, Nichols and Van Dauren, 1985) demonstrates the efficiency of the proposed method

$$
A = 10^{-3}
\begin{bmatrix}
989.50 & 5.6382 & 0.2589 & 0.0125 & 0.0006 \\
117.25 & 814.50 & 76.038 & 5.5526 & 0.3700 \\
8.7680 & 123.87 & 750.20 & 107.96 & 11.245 \\
0.9108 & 17.991 & 183.81 & 668.34 & 150.78 \\
0.0179 & 0.3172 & 1.6974 & 13.298 & 985.19
\end{bmatrix}
$$

$$
B^T = 10^{-3}
\begin{bmatrix}
0.0192 & 6.0733 & 8.2911 & 9.1965 & 0.7025 \\
-0.0013 & -0.6192 & -13.339 & -18.442 & -1.4252
\end{bmatrix}
$$

$$Q = I_5 . \qquad R = I_2 .$$

These matrices are obtained from (Kautsky, Nichols and Van Dauren, 1985) by performing a discretization with the sampling rate $\Delta T = 0.1$. The small weakly coupling parameter ε is built in the problem. It can be roughly estimated from the strongest coupled matrix - in this case matrix B. Apparently the strongest coupling is in the third row, that is

$$\varepsilon = \frac{b_{31}}{b_{32}} = \frac{8.2911}{13.339} \approx 0.62$$

The simulation results are presented in Table 7.1.

Note that the zeroth-order solution ($j = 0$) and the required solution ($j = 14$) are pretty far apart (that means, the real ε built in this problem is either bigger than 0.62 or the constant K in the definition of $O(\varepsilon^{2j}) < K\varepsilon^{2j}$ is very big) so that it takes 14 iterations to achieve desired accuracy of 10^{-3}.

In summary, the reduced order optimal solution of the algebraic discrete weakly coupled Riccati equation is obtained. This results will play the important role in the design procedure of the optimal and near-optimal controllers and filters for weakly coupled discrete systems since it reduces off-line computational requirements.

j	$P_1^{(j)}$	$P_2^{(j)}$	$P_3^{(j)}$
0	95.245 4.1053 3.0264	18.491 14.867 198.35 6.1902 3.4164 22.577	3.3032 1.3547 6.0305 2.5004 6.6735 103.77
1	177.70 11.344 4.6362	35.194 25.521 317.25 9.1415 4.9932 36.006	4.3012 1.8746 9.9935 2.7705 8.8997 124.83
2	236.17 16.951 5.4580	45.819 31.412 372.20 10.613 5.7730 42.771	4.7748 2.1239 12.113 2.9003 9.9917 133.82
4	285.96 22.098 6.0955	54.651 35.842 406.63 11.703 6.3237 47.150	5.1155 2.2973 13.515 2.9858 10.637 138.38
6	298.56 23.458 6.2522	56.866 36.884 413.56 11.960 6.4460 47.998	5.1942 2.3351 13.783 3.0035 10.756 138.59
8	301.55 23.782 6.2887	57.387 37.122 415.05 12.018 6.4731 48.170	5.2120 2.3434 13.836 3.0073 10.779 138.72
10	302.24 23.857 6.2970	57.506 37.177 415.38 12.031 6.4791 48.207	5.2160 2.3452 13.847 3.0082 10.784 138.75
12	302.40 23.874 6.2989	57.533 37.189 415.45 12.043 6.4804 48.212	5.2169 2.3456 13.850 3.0083 10.786 138.76
14	302.43 23.878 6.2993	57.539 37.192 415.47 12.035 6.4808 48.217	5.2171 2.3457 13.850 3.0084 10.786 138.76
	$P_1 = P_1^{(14)}$	$P_2 = P_2^{(14)}$	$P_3 = P_3^{(14)}$

Table 7.1 Solution of the discrete algebraic Riccati equation

7.2 Recursive Reduced-Order Solution to the Stochastic Linear Weakly Coupled Discrete Systems

The linear-quadratic Gaussian control problem of weakly coupled discrete systems has not been studied in the literature yet. Up to our best knowledge, the discrete linear weakly coupled systems has been studied so far only in (Shen and Gajić, 1989b,c). This is due to the fact that the partitioned form of the main equation of the optimal linear control theory - Riccati equation, has a very complicated from in the discrete time domain. In the previous section that problem is overcome by the use of the bilinear transformation which is applicable under quite mild assumption, so that the reduced-order solution of the discrete algebraic Riccati equation of weakly coupled systems is obtained up to any order of accuracy, by using known reduced-order results for the corresponding continuous-time algebraic Riccati equation.

Although the duality of the filter Riccati equation and regulator Riccati equation can be used together with results reported in (Shen and Gajić, 1989b) to obtain corresponding approximations to the regulator and filter gains, such approximations will not be sufficient because they only reduce the off-line computations of implementing the Kalman filter which will be of the same order as the overall weakly coupled system. The weakly coupled structure of the global Kalman filter is exploited in this section such that it may be replaced by two lower order local filters. This has been achieved via the use of the decoupling transformation introduced in (Gajić and Shen, 1989a).

The decoupling transformation of (Gajić and Shen, 1989a) is used for the exact block diagonalization of the global Kalman filter. The approximate feedback control law is then obtained by approximating the coefficients of the optimal local filters with the accuracy of $O(\epsilon^N)$. The resulting feedback control law is shown to be a near optimal solution of the LQG by studying the corresponding closed loop system as a system driven by white noise. It is shown that the order of approximation of the optimal performance is $O(\epsilon^N)$, and the order of approximation of the optimal system trajectories is $O(\epsilon^{2N})$. All required coefficients of desired accuracy are easily obtained by using the recursive reduced-order fixed point type numerical technique developed in Chapter 2. Obtained numerical algorithms converge to the required optimal coefficients with the rate of convergence of $O(\varepsilon^2)$. In addition, only low-order subsystems are involved in the algebraic computations and no analyticity requirements are imposed on the system coefficients - which is the standard assumption in the power series expansion method. As a consequence of these, under very mild conditions (coefficients are bounded functions of a small coupling parameter), in addition to the standard stabilizability-detectability subsystem assumptions, we have achieved the reduction in both off-line and on-line computational requirements.

A real world example, a fifth order distillation column, is included in this section in order to demonstrate the efficiency of the proposed method.

7.2.1 Linear Quadratic Gaussian Control of Discrete Weakly Coupled Systems at Steady State

Consider a discrete linear weakly coupled stochastic system:

$$x_1(n+1) = A_{11}x_1(n) + \varepsilon A_{12} x_2(n) + B_{11}u_1(n) + \varepsilon B_{12}u_2(n)$$
$$+ G_{11}w_1(n) + \varepsilon G_{12}w_2(n) \tag{7.5a}$$

$$x_2(n+1) = \varepsilon A_{21}x_1(n) + A_{22} x_2(n) + \varepsilon B_{21}u_1(n) + B_{22}u_2(n)$$
$$+ \varepsilon G_{21}w_1(n) + G_{22}w_2(n) \tag{7.5b}$$

$$y_1(n) = C_{11}x_1(n) + \varepsilon C_{12}x_2(n) + v_1(n) \tag{7.6a}$$

$$y_2(n) = \varepsilon C_{21}x_1(n) + C_{22}x_2(n) + v_2(n) \tag{7.6b}$$

with the performance criterion

$$J = 1/2E\left\{\sum_{n=0}^{\infty}\left[z^T(n)z(n) + u_1^T(n)R_1 u_1(n) + u_2^T(n)R_2 u_2(n)\right]\right\} \tag{7.7}$$

where $x_i \in R^{n_i}$, comprise state vectors, $u_i \in R^{m_i}$ are control inputs, $y_i \in R^{l_i}$ are observed outputs, $w_i \in R^{r_i}$ and $v_i \in R^{l_i}$ are independent zero-mean stationary Gaussian mutually uncorrelated white noise processes with intensities $W_i > 0$ and $V_i > 0$, respectively, and $z_i \in R^{s_i}$, $i = 1, 2$ are controlled outputs given by

$$z_1(n) = D_{11}x_1(n) + \varepsilon D_{12} x_2(n) \tag{7.8a}$$

$$z_2(n) = \varepsilon D_{21} x_1(n) + D_{22} x_2(n) \tag{7.8b}$$

All matrices are bounded functions of a small coupling parameter ε, having appropriate dimensions. In addition, it is assumed that R_i, $i = 1, 2$, are positive definite matrices.

The optimal control law is given by (Kwakernaak and Sivan, 1972)

$$u(n) = - F\hat{x}(n) \tag{7.9}$$

with

$$\hat{x}(n+1) = A\hat{x}(n) + Bu(n) + K\left[y(n) - C\hat{x}(n)\right] \tag{7.10}$$

where

$$A = \begin{bmatrix} A_{11} & \varepsilon A_{12} \\ \varepsilon A_{21} & A_{22} \end{bmatrix}, \quad B = \begin{bmatrix} B_{11} & \varepsilon B_{12} \\ \varepsilon B_{21} & B_{22} \end{bmatrix}, \quad C = \begin{bmatrix} C_{11} & \varepsilon C_{12} \\ \varepsilon C_{21} & C_{22} \end{bmatrix}$$

$$K = \begin{bmatrix} K_{11} & \varepsilon K_{12} \\ \varepsilon K_{21} & K_{22} \end{bmatrix}, \quad F = \begin{bmatrix} F_{11} & \varepsilon F_{12} \\ \varepsilon F_{21} & F_{22} \end{bmatrix}$$

The regulator gain F and the filter gain K are obtained from

$$F = (R + B^TPB)^{-1}B^TPA \tag{7.11}$$

$$K = AQC^T(V + CQC^T)^{-1} \tag{7.12}$$

where P and Q are positive semi-definite stabilizing solutions of the discrete time algebraic regulator and filter Riccati equations respectively given by

$$P = D^T D + A^T PA - A^T PB (R + B^T PB)^{-1} B^T PA \tag{7.13}$$

$$Q = AQA^T - AQC^T (V + CQC^T)^{-1} CQA^T + GWG^T \tag{7.14}$$

with

$$R = \text{diag}(R_1, R_2) \qquad W = \text{diag}(W_1, W_2) \qquad V = \text{diag}(V_1, V_2)$$

and

$$D = \begin{bmatrix} D_{11} & \varepsilon D_{12} \\ \varepsilon D_{21} & D_{22} \end{bmatrix}, \qquad G = \begin{bmatrix} G_{11} & \varepsilon G_{12} \\ \varepsilon G_{21} & G_{22} \end{bmatrix}$$

Due to block dominant structure of the problem matrices the required solutions P and Q have the form

$$P = \begin{bmatrix} P_{11} & \varepsilon P_{12} \\ \varepsilon P_{12}^T & P_{22} \end{bmatrix}, \qquad Q = \begin{bmatrix} Q_{11} & \varepsilon Q_{12} \\ \varepsilon Q_{12}^T & Q_{22} \end{bmatrix} \tag{7.15}$$

In order to obtain required solutions of (7.13)-(7.14) in terms of the reduced-order problems and overcome the complicated partitioned form of the discrete-time algebraic Riccati equation, we have used the method developed in the previous section, to transform the discrete Riccati equations (7.13)-(7.14) into continuous-time algebraic Riccati equations of the form

$$A_R^T P + PA_R - PS_R P + D_R^T D_R = 0, \qquad S_R = B_R R_R^{-1} B_R^T \tag{7.16}$$

$$A_F Q + QA_F^T - QS_F Q + G_F W_F G_F^T = 0, \qquad S_F = C_F^T V_F^{-1} C_F \tag{7.17}$$

such that the solutions of (7.13)-(7.14) are equal to the solutions of (7.16) and (7.17), i.e.

$$P = P, \qquad Q = Q \tag{7.18}$$

where

$$A_R = I - 2(\Delta_R^{-1})^T$$

$$B_R R_R^{-1} B_R^T = 2(I + A)^{-1} B R^{-1} B^T \Delta_R^{-1}$$

$$D_R^T D_R = 2\Delta_R^{-1} D^T D (I + A)^{-1} \tag{7.19a}$$

$$\Delta_R = (I + A^T) + D^T D (I + A)^{-1} B R^{-1} B^T$$

and

$$A_F = I - 2(\Delta_F^{-1})$$

$$C_F^T V_F^{-1} C_F = 2(I + A^T)^{-1} C^T V^{-1} C \Delta_F^{-1}$$

$$G_F W_F G_F^T = 2\Delta_F^{-1} G W G^T (I + A^T)^{-1} \tag{7.19b}$$

$$\Delta_F = (I + A) + G W G^T (I + A^T)^{-1} C^T V^{-1} C$$

It is shown in Section 7.1 that the equations (7.16)-(7.17) preserve the structure of weakly coupled systems. These equations can be solved in terms of the reduced-order problems very efficiently by using the recursive method developed in Section 2.3.2, which converges with the rate of convergence of $O(\varepsilon^2)$. Solutions of (7.16) and (7.17) are found in terms of the reduced-order problems by imposing standard stabilizability and detectability assumptions on subsystems (see Section 2.3.2).

Getting approximate solutions for P and Q in terms of the reduced-order problems will produce savings in off-line computations. However, in the case of stochastic systems, where an additional system - filter - has to be built, one is particularly interested in the reduction of on-line computations. In this section that will be achieved by the use of the decoupling transformation introduced in (Gajić and Shen, 1989a). The basic properties of that transformation in the discrete-time domain are given in Appendix 7.1.

The Kalman filter (7.10) is viewed as a system driven by the innovation process. However, one might study the filter form when it is driven by both measurements and controls. The filter form under consideration is obtained from (7.10) as

$$\hat{x}_1(n+1) = (A_{11} - B_{11}F_{11} - \epsilon^2 B_{12} F_{12})\hat{x}_1(n) + \epsilon(A_{12} - B_{11}F_{12} - B_{12} F_{22})\hat{x}_2(n)$$

$$+ K_{11}v_1(n) + \epsilon K_{12}v_2(n) \tag{7.20a}$$

$$\hat{x}_2(n+1) = \epsilon(A_{21} - B_{21}F_{11} - B_{22}F_{21})\hat{x}_1(n) + (A_{22} - \epsilon^2 B_{21}F_{12} - B_{22}F_{22})\hat{x}_2(n)$$

$$+ \epsilon K_{21}v_1(n) + K_{22}v_2(n) \tag{7.20b}$$

with innovation processes

$$v_1(n) = y_1(n) - C_{11}\hat{x}_1(n) - \epsilon C_{12}\hat{x}_2(n) \tag{7.21.a}$$

$$v_2(n) = y_2(n) - \epsilon C_{21}\hat{x}_1(n) - C_{22}\hat{x}_2(n) \tag{7.21.b}$$

The nonsingular state transformation of (Gajić and Shen, 1989a) will block diagonalize (7.20) under condition that $(A_{11} - B_{11}F_{11} - \epsilon^2 B_{12}F_{21})$ and $(A_{22} - B_{22}F_{22} - \epsilon^2 B_{21}F_{12})$ have no eigenvalues in common (see Appendix 7.2). This transformation is given by

$$\begin{bmatrix} \hat{\eta}_1(n) \\ \hat{\eta}_2(n) \end{bmatrix} = \begin{bmatrix} I - \epsilon^2 LH & -\epsilon L \\ \epsilon H & I \end{bmatrix} \begin{bmatrix} \hat{x}_1(n) \\ \hat{x}_2(n) \end{bmatrix} = T \begin{bmatrix} \hat{x}_1(n) \\ \hat{x}_2(n) \end{bmatrix}$$

with

$$T^{-1} = \begin{bmatrix} I & \epsilon L \\ -\epsilon H & I - \epsilon^2 HL \end{bmatrix} \tag{7.22}$$

where matrices L and H satisfy equations

$$Ha_{11} - a_{22}H + a_{21} - \epsilon^2 Ha_{12}H = 0 \tag{7.23}$$

$$L(a_{22} + \epsilon Ha_{12}) - (a_{11} - \epsilon^2 a_{12}H)L - a_{12} = 0 \tag{7.24}$$

with

$$a_{11} = A_{11} - B_{11}F_{11} - \varepsilon^2 B_{12} F_{12}$$

$$a_{12} = A_{12} - B_{11}F_{12} - B_{12} F_{22}$$

$$a_{21} = A_{21} - B_{21} F_{11} - B_{22} F_{21}$$

$$a_{22} = A_{22} - B_{22} F_{22} - \varepsilon^2 B_{21} F_{12}$$

The optimal feedback control, expressed in the new coordinates, has the form

$$u_1(n) = -f_{11}\hat{\eta}_1(n) - \varepsilon f_{12}\hat{\eta}_2(n) \tag{7.25a}$$

$$u_2(n) = -\varepsilon f_{21}\hat{\eta}_1(n) - f_{22}\hat{\eta}_2(n) \tag{7.25b}$$

with

$$\hat{\eta}_1(n+1) = \alpha_1\hat{\eta}_1(n) + \beta_{11}v_1(n) + \varepsilon\beta_{12}v_2(n) \tag{7.26a}$$

$$\hat{\eta}_2(n+1) = \alpha_1\hat{\eta}_2(n) + \varepsilon\beta_{21}v_1 + \beta_{22}v_2(n) \tag{7.26b}$$

where

$$f_{11} = F_{11} - \varepsilon^2 F_{12}H, \qquad f_{12} = F_{12} + (F_{11} - \varepsilon^2 F_{12} H)L$$

$$f_{21} = F_{21} - F_{22} H, \qquad f_{22} = F_{22} + \varepsilon^2(F_{21} - F_{22}H)L$$

$$\alpha_1 = a_{11} - \varepsilon^2 a_{12} H, \qquad \alpha_2 = a_{22} + \varepsilon^2 Ha_{12}$$

$$\beta_{11} = K_{11} - \varepsilon^2 L(H + K_{21}), \qquad \beta_{12} = K_{12} - KL_{22} - \varepsilon^2 LHK_{12}$$

$$\beta_{21} = HK_{11} + K_{21}, \qquad \beta_{22} = K_{22} + \varepsilon^2 HK_{12}$$

The innovation processes v_1 and v_2 are now given by

$$v_1(n) = y_1(n) - d_{11}\hat{\eta}_1(n) - \varepsilon d_{12}\hat{\eta}_2(n) \tag{7.27a}$$

$$v_2(n) = y_2(n) - \varepsilon d_{21}\hat{\eta}_1(n) - d_{22}\hat{\eta}_2(n) \tag{7.27b}$$

where

$$d_{11} = C_{11} - \varepsilon^2 C_{12} H . \qquad d_{12} = C_{11}L + C_{12} - \varepsilon^2 C_{12} HL$$

$$d_{21} = C_{21} - C_{22} H . \qquad d_{22} = C_{22} + \varepsilon^2(C_{21} - C_{22} H)L$$

Approximate control laws are defined by perturbing coefficients F_{ij}, K_{ij} (i, j = 1, 2), L and H by $O(\varepsilon^k)$, k = 1, 2, ..., in other words by using k-th approximations for these coefficients, where k stands for the required order of accuracy, that is

$$u_1^{(k)}(n) = -f_{11}^{(k)}\hat{\eta}_1^{(k)}(n) - \varepsilon f_{12}^{(k)}\hat{\eta}_2^{(k)}(n) \tag{7.28a}$$

$$u_2^{(k)}(n) = -\varepsilon f_{21}^{(k)}\hat{\eta}_1^{(k)}(n) - f_{22}^{(k)}\hat{\eta}_2^{(k)}(n) \tag{7.28b}$$

with

$$\hat{\eta}_1^{(k)}(n+1) = \alpha_1^{(k)}\hat{\eta}_1^{(k)}(n) + \beta_{11}^{(k)}v_1^{(k)}(n) + \varepsilon\beta_{12}^{(k)}v_2^{(k)}(n) \tag{7.29a}$$

$$\hat{\eta}_2^{(k)}(n+1) = \alpha_2^{(k)}\hat{\eta}_2^{(k)}(n) + \varepsilon\beta_{21}^{(k)}v_1^{(k)}(n) + \beta_{22.}^{(k)}v_2^{(k)}(n) \tag{7.29b}$$

where

$$v_1^{(k)}(n) = y_1^{(k)}(n) - d_{11}^{(k)}\hat{\eta}_1^{(k)}(n) - \varepsilon d_{12}^{(k)}\hat{\eta}_2^{(k)}(n) \tag{7.30a}$$

$$v_2^{(k)}(n) = y_2^{(k)}(n) - \varepsilon d_{21}^{(k)}\hat{\eta}_1^{(k)}(n) - d_{22}^{(k)}\hat{\eta}_2^{(k)}(n) \tag{7.30b}$$

and

$$f_{ij}^{(k)} = f_{ij} + O(\varepsilon^k). \qquad d_{ij}^{(k)} = d_{ij} + O(\varepsilon^k)$$

$$\beta_{ij}^{(k)} = \beta_{ij} + O(\varepsilon^k). \qquad \alpha_{ij}^{(k)} = \alpha_{ij} + (\varepsilon^k) \qquad i, j = 1, 2.$$

The approximate values of $J^{(k)}$ are obtained from the following equations

$$J^{(k)} = 1/2E\left\{\sum_{n=0}^{\infty}\left[x^{(k)^T}(n)\,D^TDx^{(k)}(n) + u^{(k)}\,(\hat{n})Ru^{(k)}(n)\right]\right\}$$

$$= 1/2\mathrm{tr}\left\{D^TDq_{11}^{(k)} + f^{(k)^T}Rf^{(k)}\,q_{22}^{(k)}\right\} \qquad (7.31)$$

where

$$q_{11}^{(k)} = \mathrm{Var}\left\{(x_1^{(k)}\ x_2^{(k)})^T\right\} \text{ and } q_{22}^{(k)} = \mathrm{Var}\left\{(\hat{n}_1^{(k)}\ \hat{n}_2^{(k)})^T\right\}$$

$$u^{(k)} = \begin{bmatrix} u_1^{(k)}(n) \\[4pt] u_2^{(k)}(n) \end{bmatrix}, \qquad f^{(k)} = \begin{bmatrix} f_{11}^{(k)} & \varepsilon f_{12}^{(k)} \\[4pt] \varepsilon f_{21}^{(k)} & f_{22}^{(k)} \end{bmatrix}$$

Quantities $q_{11}^{(k)}$ and $q_{22}^{(k)}$ can be obtained by studying the variance equation of the following system driven by white noise

$$\begin{bmatrix} x^{(k)}(n+1) \\[6pt] \hat{\eta}^{(k)}(n+1) \end{bmatrix} = \begin{bmatrix} A & -Bf^{(k)} \\[6pt] \beta^{(k)}C & \alpha^{(k)} - \beta^{(k)}d^{(k)} \end{bmatrix} \begin{bmatrix} x^{(k)}(n) \\[6pt] \hat{\eta}^{(k)}(n) \end{bmatrix}$$

$$(7.32)$$

$$+ \begin{bmatrix} G & O \\[6pt] O & \beta^{(k)} \end{bmatrix} \begin{bmatrix} w(n) \\[6pt] v(n) \end{bmatrix}$$

where

$$\alpha^{(k)} = \begin{bmatrix} \alpha_1^{(k)} & 0 \\[4pt] 0 & \alpha_2^{(k)} \end{bmatrix}, \ \beta^{(k)} = \begin{bmatrix} \beta_{11}^{(k)} & \varepsilon\beta_{12}^{(k)} \\[4pt] \varepsilon\beta_{21}^{(k)} & \beta_{22}^{(k)} \end{bmatrix}, \ d^{(k)} = \begin{bmatrix} d_{11}^{(k)} & \varepsilon d_{12}^{(k)} \\[4pt] \varepsilon d_{21}^{(k)} & d_{22}^{(k)} \end{bmatrix}$$

Equation (7.32) can be represented in a composite form,

$$r^{(k)}(n+1) = \Lambda^{(k)}\,r^{(k)}(n) + \Pi^{(k)}\,\omega(n) \qquad (7.33)$$

with obvious definition for $\Lambda^{(k)}$, $\Pi^{(k)}$, $r^{(k)}(n)$ and $\omega(n)$. The variance of $r^{(k)}$ (n), at steady state denoted by $q^{(k)}$, is given by the discrete algebraic Lyapunov equation (Kwakernaak and Sivan, 1972)

$$q^{(k)} = \Lambda^{(k)}\,q^{(k)}\Lambda^{(k)^T} + \Pi^{(k)}\,W\Pi^{(k)^T}, \qquad W = \mathrm{diag}\ (W,V) \qquad (7.34)$$

with $q^{(k)}$ partitioned as

$$q^{(k)} = \begin{bmatrix} q_{11}^{(k)} & q_{12}^{(k)} \\ q_{12}^{(k)T} & q_{22}^{(k)} \end{bmatrix} \qquad (7.35)$$

On the other hand, the optimal value of J has the very well known from, (Kwakernaak and Sivan, 1972)

$$J^{opt} = 1/2\mathrm{tr}\left[D^T DQ + PK(CQC^T + V)K^T\right] \qquad (7.36)$$

where P, Q, F and K are obtained from (7.11)–(7.14).

The near-optimality of the proposed approximate control law (7.28) is established in the following theorem.

Theorem 7.1 Let x_1 and x_2 be the optimal trajectories and J be the optimal value of the performance criterion. Let $x_1^{(k)}$, $x_2^{(k)}$ and $J^{(k)}$ be corresponding quantities under the approximate control law $u^{(k)}$ given by (7.28). Then under the condition stated in Lemma 7.1 and under the standard stabilizability and detectability subsystem assumptions the following hold

$$J^{opt} - J^{(k)} = O(\varepsilon^k) \qquad (7.37a)$$

$$\mathrm{Var}(x_1 - x_1^{(k)}) = O(\varepsilon^{2k}), \quad k = 0, 1, 2, \ldots . \qquad (7.37b)$$

The proof of this theorem is rather lenghtly and it is omitted. It follows ideas of Theorems 1 and 2 from (Khalil an Gajić, 1984) obtained for the another class of small parameter problems - singularly perturbed systems. These two theorems were proved in the context of weakly coupled linear systems in (Shen and Gajić, 1989a). In addition, due to the discrete nature of the problem, the proof of our theorem 7.1, utilizes the bilinear transformation from (Power, 1967) which transforms the discrete Lyapunov equation into the continuous one and compare it with the corresponding equation under the optimal control law. More about it can be found in (Shen, 1989).

7.2.2 Numerical Example

A real world physical example (a fifth order distillation column) studied in Section 7.1.2 demonstrates the efficiency of the proposed method. The problem matrices A and B are given in Section 7.1.2. Remaining matrices are chosen as

$$C = \begin{bmatrix} 1 & 1 & 0 & 0 & 0 \\ 0 & 0 & 1 & 1 & 1 \end{bmatrix}, \qquad Q = I_5, \qquad R = I_2$$

It is assumed that G = B, and that white noise intensity matrices are given by

$$W_1 = 1, \qquad W_2 = 1, \qquad V_1 = 0.1 \qquad V_2 = 0.1$$

The simulation results are presented in the following table

k	$J^{(k)}$	$J^{(k)} - J$
0	0.80528×10^{-2}	0.6989×10^{-3}
1	0.75977×10^{-2}	0.2438×10^{-3}
2	0.74277×10^{-2}	0.7380×10^{-4}
4	0.73887×10^{-2}	0.3480×10^{-4}
6	0.73546×10^{-2}	0.5000×10^{-6}
8	0.73539×10^{-2}	$< 10^{-7}$
optimal	0.73539×10^{-2}	

Table 7.2 Approximate values for criterion

It can be seen, that despite of the relatively big value of the coupling parameter $\varepsilon = 0.62$, we have very rapid convergence to the optimal solution.

In summary, the near-optimum (up to any desired accuracy) steady state regulators are obtained for the stochastic linear weakly coupled discrete systems. The proposed method reduces considerably the size of required off-line and on-line computations since it introduces the full parallelism in the design procedure.

Appendix 7.1

Block Diagonalization of Weakly Coupled
Linear Discrete Systems

Consider a weakly coupled discrete system

$$x_1(n+1) = A_1 x_1(n) + \varepsilon A_2 x_2(n)$$

$$x_2(n+1) = \varepsilon A_3 x_1(n) + A_4 x_2(n)$$

(a.1)

It can be shown, by forming the discrete version of (Gajić and Shen, 1989a), that the nonsingular transformation

$$T = \begin{bmatrix} I - \varepsilon^2 LH & -\varepsilon L \\ \varepsilon H & I \end{bmatrix}, \quad T^{-1} = \begin{bmatrix} I & \varepsilon L \\ -\varepsilon H & I - \varepsilon^2 HL \end{bmatrix}$$

(a.2)

will transform (a.1) into a block diagonal form

$$\eta_1(n+1) = (A_1 - \varepsilon^2 A_2 H)\eta_1(n)$$

$$\eta_2(n+1) = (A_4 - \varepsilon^2 HA_2)\eta_2(n)$$

(a.3)

The transformation matrices L and H satisfy algebraic equations

$$HA_1 - A_4 H + A_3 - \varepsilon^2 HA_2 H = 0$$

(a.4)

$$L(A_4 + \varepsilon^2 HA_2) - (A_1 - \varepsilon^2 A_2 H)L - A_2 = 0$$

(a.5)

The following lemma is obtained in (Gajić and Shen, 1989a):

Lemma 7.1 Under assumption that matrices A_1 and A_4 have no eingenvalues is common there exists a small parameter ε such that the unique solutions of (a.4) and (a.5) exist.

In our problem matrices A_1 and A_4 are feedback matrices so that the assumption of Theorem 7.1 is almost always satisfied.

Numerical solution of (a.4) can be obtained by using a fixed point type algorithm (Gajić and Shen, 1989a)

$$H^{(j+1)}A_1 - A_4 H^{(j+1)} + A_3 - \varepsilon^2 H^{(j)} A_2 H^{(j)} = 0, \quad j = 0, 1, 2, ..., N-1. \quad (a.6)$$

where $H^{(0)}$ is obtained from

$$H^{(0)}A_1 - A_4 H^{(0)} + A_3 = 0$$

This algorithm has the rate of convergence of $O(\varepsilon^2)$. Having obtained $H^{(N)}$, the equation (a.5) can be solved as a linear Sylvester type equation. Another way of solving (a.4) exploits the Newton method developed in (Gajić and Shen, 1989a).

LINEAR DISCRETE SINGULARLY PERTURBED CONTROL SYSTEMS

8.1 Recursive Solution to the Discrete Linear Quadratic Control Problem of Singularly Perturbed Systems

8.1.1 Introduction

The linear singularly perturbed discrete systems have been studied recently in different set-ups by many researchers. Two main structures of singularly perturbed linear discrete systems were considered: the fast time scale version (Butuzov and Vasileva, 1971, Hoppensteadt and Miranker 1977, Blankenship, 1981, Litkouhi, 1983, Litkouhi and Khalil, 1984, 1985, Mahmoud 1986, Oloomi and Sawan 1987, Khorasani and Azimi-Sadjadi, 1987.) and the slow time scale version (e.g. Phillips, 1980, Naidu and Rao, 1985). Discrete time models of the singularly perturbed linear systems, similar to (Phillips, 1980, Naidu and Rao 1985), are studied by Mahmoud and his coworkers also (Mahmoud, Chen and Singh, 1986). Since the slow time scale version presupposes the asymptotic stability of the fast models, it seems that in the design procedure of stabilizing feedback controllers, the fast time scale version is much more appropriate (Litkouhi and Khalil, 1985). In this chapter, we will adopt the structure of singularly perturbed discrete linear systems defined by Litkouhi and Khalil, and study the corresponding linear-quadratic control problem. We will take a new approach, based on a bilinear transformation (Kondo and Furuta, 1986). It is known that the main equation of the optimal linear control theory - the Riccati equation, has quite complicated form in the discrete time domain. Partitioning this equation, in the sense of singular perturbation methodology, will produce a lot of terms (partitioned inversion of a matrix sum) and make corresponding problem numerically inefficient, even though the problem order-reduction is achieved. By applying a bilinear transformation, the solution of a discrete algebraic Riccati equation of singularly perturbed systems is obtained by using already known results for the corresponding continuous algebraic Riccati equation. The proposed method produces the reduced-order near-optimal solution, up to an arbitrary order of accuracy, that is of $O(\varepsilon^k)$, where ε is a small perturbation parameter. It reduces the size of required computations, and it is very suitable for parallel programming. The real world example, an F-8 aircraft demonstrates the efficiency of the method introduced.

The importance of the existence of the $O(\varepsilon^k)$ theory, for a small parameter problems, is indicated in (Gajic, Petkovski and Harkara, 1989, Shen and Gajic, 1989a). It was shown in (Gajic, Petkovski and Harkara, 1989) that the $O(\varepsilon)$ theory fails to produce required results. In (Shen and Gajic, 1989a) the approximate filter has to be obtained with an accuracy of at least $O(\varepsilon^6)$ in order to stabilize the system.

8.1.2 Reduced-Order Near-Optimal Solution of the Discrete Algebraic Riccati Equation of Singularly Perturbed Systems.

The algebraic Riccati equation of singularly perturbed linear discrete systems is given by

$$P = A^TPA + Q - A^T PB(B^TPB + R)^{-1}B^TPA, \qquad R > 0, Q \geq 0 \qquad (8.1)$$

where (Litkouhi, 1983, Litkouhi and Khalil 1984, 1985)

$$A = \begin{bmatrix} I + \varepsilon A_1 & \varepsilon A_2 \\ A_3 & A_4 \end{bmatrix}, \qquad B = \begin{bmatrix} \varepsilon B_1 \\ B_2 \end{bmatrix}, \qquad Q = \begin{bmatrix} Q_1 & Q_2 \\ Q^T_2 & Q_3 \end{bmatrix} \qquad (8.2)$$

and ε is a small positive singular perturbation parameter. In addition, the following condition is satisfied (Litkouhi and Khalil, 1985)

$$\det (I - A_4) \neq 0 \tag{8.3}$$

Due to the special structure of the problem matrices and its representation in the fast time scale, the required solution P has the form (Litkouhi and Khalil, 1984)

$$P = \begin{bmatrix} P_1/\varepsilon & P_2 \\ P_2^T & P_3 \end{bmatrix} \tag{8.4}$$

The main goal in the theory of singular perturbations is to obtain the required solution in terms of the reduced-order problems, namely subsystems. In the case of the algebraic singularly perturbed discrete Riccati equation (8.1), the inversion of the partitioned matrix $(B^T PB + R)$ will produce a lot of terms and make corresponding approach computationally very involved even though one is faced with the reduced-order numerical problems.

In order to overcome this problem, we have used the bilinear transformation introduced in (Kondo and Furuta, 1986) to transform the discrete Riccati equation (8.1) into a continuous-time algebraic Riccati equation of the form

$$A_c^T P_c + P_c A_c + Q_c - P_c S_c P_c = 0, \qquad S_c = B_c R_c^{-1} B_c^T \tag{8.5}$$

such that the solution of (8.1) is equal to the solution of (8.5).

It will be shown that the equation (8.5) preserves the structure of singularly perturbed systems. This equation can be solved in terms of the reduced-order problems very efficiently by using the recursive method developed in Chapter 2, which converges with the rate of convergence of $O(\varepsilon)$.

The bilinear transformation states that equations (8.1) and (8.5) have the same solution if the following hold (Kondo and Furuta, 1986)

$$A_c = I - 2D^{-T} \tag{8.6a}$$

$$S_c = 2(I + A)^{-1} S_d D^{-1}, \qquad S_d = BR^{-1}B^T \tag{8.6b}$$

$$O_c = 2D^{-1} Q(I + A)^{-1} \tag{8.6c}$$

$$D = (I + A^T) + Q(I + A)^{-1} S_d \tag{8.6d}$$

assuming that $(I + A)^{-1}$ exists. It can be easily seen that the matrix

$$I + A = \begin{bmatrix} 2I + \varepsilon A_1 & \varepsilon A_2 \\ A_3 & I + A_4 \end{bmatrix} \tag{8.7}$$

is invertible for small values of ε if and only if the matrix $I + A_4$ is invertible. Using the standard result from (Stewart, 1973), it means that A_4 has no eigenvalues at -1. This is a very mild constraint. Thus, the method proposed in this chapter will be applicable under the following assumption.

Assumption 8.1 The fast subsystem matrix has no eigenvalues located at -1.

It is important to point out that, under given assumption, the matrix D defined in (8.6d) is nonsingular (Bar-Ness and Halbersberg, 1980).

Let us show that the application of the bilinear transformation preserves the structure of singularly perturbed systems, namely, matrices defined in (8.6) should correspond to the linear-quadratic (LQ) singularly perturbed continuous control problem.

Using the formula for an inversion of block partitioned matrices, the following can be shown from (8.2) and (8.6)

$$(I + A)^{-1} = \begin{bmatrix} I + O(\varepsilon) & O(\varepsilon) \\ O(1) & O(1) \end{bmatrix} \tag{8.8a}$$

$$D^f = \begin{bmatrix} I + O(\varepsilon) & O(1) \\ O(\varepsilon) & O(1) \end{bmatrix}, \quad D^{fT} = \begin{bmatrix} I + O(\varepsilon) & O(\varepsilon) \\ O(1) & O(1) \end{bmatrix} \qquad (8.8b)$$

$$S_d^f = \begin{bmatrix} O(\varepsilon^2) & O(\varepsilon) \\ O(\varepsilon) & O(1) \end{bmatrix} \qquad (8.8c)$$

so that

$$A_c^f = \begin{bmatrix} O(\varepsilon) & O(\varepsilon) \\ O(1) & O(1) \end{bmatrix}, \quad Q_c^f = \begin{bmatrix} O(1) & O(1) \\ O(1) & O(1) \end{bmatrix} \qquad (8.9a)$$

$$S_c^f = \begin{bmatrix} O(\varepsilon^2) & O(\varepsilon) \\ O(\varepsilon) & O(1) \end{bmatrix} \longrightarrow B_c^f = \begin{bmatrix} O(\varepsilon) \\ O(1) \end{bmatrix} \qquad (8.9b)$$

where f indicates the fast time scale version quantities.

It is the very well known fact that the structure of matrices obtained in (8.9) corresponds to the fast time scale representation of the continuous-time singularly perturbed LQ control problem (Litkouhi and Khalil, 1984; Kokotović and Khalil, 1986; Kokotović, Khalil and O'Reilly, 1986).

Since there is no difference in the use of either the slow or fast time scale representation for the continuous-time LQ control problem of singularly perturbed systems, we will adopt the slow time scale version for this problem. Even more, it is customary to represent continuous time singularly perturbed systems by their slow time version (Kokotović and Khalil, 1986; Kokotović, Khalil and O'Reilly, 1986).

Slow time version of (8.9) can be obtained by multiplying the matrix A_c^f by $1/\varepsilon$ and matrix S_c^f by $1/\varepsilon^2$. Introducing a notation for the compatible partitions of these matrices we have

$$A_c = \begin{bmatrix} A_{11} & A_{12} \\ A_{21}/\varepsilon & A_{22}/\varepsilon \end{bmatrix} \quad , \quad S_c = \begin{bmatrix} S_{11} & S_{12}/\varepsilon \\ S_{12}^{\,T}/\varepsilon & S_{22}/\varepsilon^2 \end{bmatrix} \tag{8.10}$$

By doing this, the required solution P from (8.4) obtained now from (8.5), will be multiplied by ε, that is

$$\varepsilon P = P_c = \begin{bmatrix} P_1 & \varepsilon P_2 \\ \varepsilon P_2^{\,T} & \varepsilon P_3 \end{bmatrix} \tag{8.11}$$

Going from the fast time version to the slow time version does not change the matrix Q_c. It is partitioned as

$$Q_c = \begin{bmatrix} Q_{11} & Q_{12} \\ Q_{12}^{\,T} & Q_{22} \end{bmatrix} = Q_c^{\,l} \tag{8.12}$$

It is important to notice that partitions defined in (8.10)-(8.12) have to be performed by a computer only in the process of calculations and there is no need for the corresponding analytical expressions.

Solution of (8.5) can be found in terms of the reduced-order problems by imposing standard stabilizability-detectability assumptions on the slow and fast subsystems. The efficient recursive reduced-order algorithm for solving (8.5) is obtained in Section 2.2.2. It will be briefly summarized here taking into account the specific features of the problem under study.

First of all, we derive expressions for B_c and R_c so that the analogy between the discrete quantities (A, B, Q, R) and continuous ones (A_c, B_c, Q_c, R_c) is completed.

By definition

$$S_c^{\,l} = B_c^{\,l} R_c^{-1} B_c^{\,l\,T} \tag{8.13}$$

From (8.6b) we have

$$S_c = 2(I + A)^{-1} S_d \, D^{-1}(I + A^T)\,(I + A^T)^{-1}$$

Since

$$S_d \ D^{-1}(I + A^T) = S_d\left[(I + A^T)^{-1}D\right]^{-1} = S_d\left[I + (I + A^T)^{-1} Q(I + A)^{-1} S_d\right]^{-1} =$$

$$= B\left[R + B^T(I + A^T)^{-1}Q(I + A)^{-1}B\right]^{-1}B^T$$

(the last step in this expression is justified in (Bar-Ness and Halbersberg, 1980)), we get

$$S_c^f = 2(I + A)^{-1}B\left[R + B^T(I + A^T)^{-1}Q(I + A)^{-1}B\right]^{-1}B^T(I + A^T)^{-1} \qquad (8.14)$$

Comparing (8.13) and (8.14) we conclude

$$B_c^f = (I + A)^{-1}B = \begin{bmatrix} \varepsilon B_1^f \\ B_2^f \end{bmatrix} \qquad (8.15)$$

and

$$R_c = 0.5\left[R + B^T(I + A^T)^{-1}Q(I + A)^{-1}B\right] \qquad (8.16)$$

Note that R_c is positive definite. Slow time version of (8.15) is

$$B_c = 1/\varepsilon B_c^f = \begin{bmatrix} B_1^f \\ B_2^f/\varepsilon \end{bmatrix} \qquad (8.17)$$

The $O(\varepsilon)$ approximation of (8.5) subject to (8.10), (8.12), and (8.16)-(8.17) can be obtained from the following reduced-order algebraic equations.

$$0 = \underline{P}_1\underline{A} + \underline{A}^T\underline{P}_1 + \underline{Q} - \underline{P}_1\underline{S}\underline{P}_1, \qquad \underline{S} = B_0R_0^{-1}B_0^T \qquad (8.18a)$$

$$0 = \underline{P}_3A_{22} + A_{22}^T\underline{P}_3 + Q_{22} - \underline{P}_3S_{22}\underline{P}_3 \qquad (8.18b)$$

$$\underline{P}_2 = \underline{P}_1Z_1 - Z_2 \qquad (8.18c)$$

where newly defined matrices can be obtained easily using results from Section 2.2.2.

Unique positive semi-definite stabilizing solution of (8.18) exists under the following assumption:

Assumption 8.2: Triples $(\underline{A}, B_0, \sqrt{\underline{Q}})$, and $(A_{22}, B_2, \sqrt{Q_{22}})$ are stabilizable-detectable.

Defining the approximation errors as

$$P_i = \underline{P}_i + \varepsilon E_i, \quad i = 1, 2, 3. \tag{8.19}$$

the recursive reduced-order algorithm, with the rate of convergence of $O(\varepsilon)$, can be derived similarly to (2.34)

$$E_1^{(j+1)} D_1 + D_1^T E_1^{(j+1)} = D^T H_1^{(j)^T} + H_1^{(j)} D + D^T H_3^{(j)} D + \varepsilon H_2^{(j)} \tag{8.20a}$$

$$E_2^{(j+1)} D_3 + E_1^{(j+1)} D_{21} + D_{22}^T E_3^{(j+1)} = H_1^{(j, j+1)} \tag{8.20b}$$

$$E_3^{(j+1)} D_3 + D_3^T E_3^{(j+1)} = H_3^{(j)} \tag{8.20c}$$

with $j = 0, 1, 2, ...,$ and $E_1^{(0)} = 0, E_2^{(0)} = 0, E_3^{(0)} = 0,$ where newly defined matrices are given in Appendix 8.1. It is important to point out that D_1 and D_3 are stable matrices (Gajic, 1986).

The rate of convergence of (8.20) is $O(\varepsilon)$, that is

$$\| P_i - P_i^{(j)} \| = O(\varepsilon^j), \qquad i = 1, 2, 3, \ j = 0, 1, 2, ... \ . \tag{8.21}$$

where

$$P_i^{(j)} = \underline{P}_i + \varepsilon E_i^{(j)} \qquad i = 1, 2, 3, \quad j = 0, 1, 2, ... \ . \tag{8.22}$$

In summary, the proposed algorithm for the reduced-order solution of the singularly perturbed discrete algebraic Riccati equation has the following form:

1) Transform (8.1) into (8.5) by using the bilinear transformation defined in (8.6).

2) Solve (8.5) by using the recursive reduced-order algorithm defined by (8.18)-(8.22).

It is important to point out that (8.18)-(8.22) can be implemented by using ideas from parallel programming ((8.20a) and (8.20c) can be solved simultaneously) which will play an important role in the real time control implementation.

8.2 Near-Optimal Control of Linear Singularly Perturbed Discrete Systems

The positive semi-definite stabilizing solution of the algebraic discrete Riccati equation (8.1), produces the answer to the optimal linear-quadratic steady state control problem. Namely, a quadratic criterion

$$J = \sum_{k=0}^{\infty} (x^T(k)Qx(k) + u^T(k)Ru(k)) \tag{8.23}$$

is minimized along trajectories of a linear dynamic discrete system

$$x(k+1) = Ax(k) + Bu(k) \tag{8.24}$$

by using the control input of the form

$$u(k) = - (R + B^TPB)^{-1}B^TPAx(k) \tag{8.25}$$

where P is obtained from (8.1), (Dorato and Levis, 1971). This problem has been studied in the context of singular perturbations in (Litkouhi and Khalil, 1984), where the fast time version has been adopted, so that (8.23) is multiplied by a small perturbation parameter, that is,

$$J_f = \varepsilon J \tag{8.26}$$

It is proved in (Litkouhi and Khalil, 1985) that the near-optimal control given by

$$u^{(j)}(k) = - (R + B^TP^{(j)}B)^{-1}B^TP^{(j)}Ax(k) = - F^{(j)}x(k) \tag{8.27}$$

where $P^{(j)}$ satisfies

$$P^{(j)} - P^{opt} = O(\varepsilon^j) \tag{8.28}$$

is near-optimal in the sense

$$J_\ell^{(j)} - J_\ell^{opt} = O(\varepsilon^{2j}) \qquad (8.29)$$

The approximate performance $J^{(j)}$ can be obtained from the discrete algebraic Lyapunov equation

$$K^{(j)} = (A - BF^{(j)})^T K^{(j)} (A - BF^{(j)}) + Q + F^{(j)^T} RF^{(j)} \qquad (8.30)$$

so that

$$J^{(j)} = x(0)^T K^{(j)} x(0) \qquad (8.31)$$

In the previous section we have developed a very efficient technique for generating $P^{(j)}$ by using the recursive reduced-order scheme (8.18)-(8.22), such that each iteration improves the accuracy by an order of magnitude (see (8.21)). Thus, the proposed algorithm from Section 8.1 and the theoretical results obtained in (Litkouhi and Khalil, 1985) and given in (8.27)-(8.29) comprise a new method for solving the linear-quadratic control problem of singularly perturbed discrete systems.

The efficiency of this method is demonstrated on a real world example in the next section.

8.2.1 Case Study: Discrete Model of F-8 Aircraft

Linearized model of the F-8 aircraft is considered in (Elliott, 1977). By a proper scaling this model was presented in the singularly perturbed continuous form (fast time version) in (Litkouhi, 1983), with the system matrix

$$\begin{bmatrix} -0.015 & -0.0805 & -0.0011666 & 0 \\ 0 & 0 & 0 & 0.03333 \\ -2.28 & 0 & -0.84 & 1 \\ 0.6 & 0 & -4.8 & -0.49 \end{bmatrix}$$

and with the control matrix

$$\begin{bmatrix} -0.0000916 & 0.0007416 \\ 0 & 0 \\ -0.11 & 0 \\ -8.7 & 0 \end{bmatrix}$$

Small elements in the first two rows indicate two slow variables in contrary to relatively big elements in the third and forth rows corresponding to fast variables. Small perturbation parameter ε is chosen as $\varepsilon = 1/30$. This model is discretized in (Litkouhi, 1983) by using the sampling period $T = 1$, leading to

$$A = \begin{bmatrix} 0.98475 & -0.079903 & 0.0009054 & -0.0010765 \\ 0.041588 & 0.99899 & -0.035855 & 0.012684 \\ -0.54662 & 0.044916 & -0.32991 & 0.19318 \\ 2.6624 & -0.10045 & -0.92455 & -0.26325 \end{bmatrix}$$

$$B = \begin{bmatrix} 0.0037112 & 0.00073610 \\ -0.087051 & 0.0000093411 \\ -1.19844 & -0.00041378 \\ -3.1927 & 0.00092535 \end{bmatrix}$$

The linear-quadratic control problem is solved for weighting matrices

$$R = I_2 , \quad Q = 10^{-2} I_4$$

and the initial condition $x(0) = \begin{bmatrix} 1, 0, 0.008, 0 \end{bmatrix}^T$.

The eigenvalues of the matrix A_4 are $-0.297 \pm j0.442$, so that Assumption 8.1 is satisfied.

Simulation results for the reduced-order solution of the corresponding discrete algebraic Riccati equation are presented in Table 8.1. The approximate values of the criterion are presented in Table 8.2.

j	P_1		P_2		P_3	
0	1.98280	0.00737	-0.36273	-0.30229	0.61994	0.03821
	2.21170		-1.95280	0.37275		0.34262
1	2.12850	-0.00249	-0.40041	-0.31446	0.67749	0.02734
	2.34370		-2.07120	0.40444		0.34488
2	2.13680	-0.00359	-0.40292	-0.31533	0.68128	0.02656
	2.35100		-2.07690	0.40631		0.34514
3	2.13700	-0.00368	-0.40304	-0.31537	0.68147	0.02652
	2.35110		-2.07700	0.40637		0.34516
4	2.13700	-0.00368	-0.40304	-0.31537	0.68147	0.02651
	2.35110		-2.07700	0.40637		0.34516
*	2.13700	-0.00368	-0.40304	-0.31537	0.68147	0.02651
	2.35110		-2.07700	0.40637		0.34516

Table 8.1 Reduced-order solution of the discrete algebraic Riccati equation
(* --- exact solution)

j	$J_{apr}^{(j)} - J_{opt}$
0	0.208×10^{-2}
1	0.885×10^{-5}
2	0.155×10^{-7}
3	0.534×10^{-10}

Table 8.2 Near-optimality of the approximate criterion

8.3 Parallel Reduced-Order Controllers for Stochastic Linear Singularly Perturbed Systems

8.3.1 Introduction

The continuous time LQG problem of singularly perturbed systems is solved in (Khalil and Gajic, 1984) by using the power series expansion approach, and later on in (Gajic, 1986) by using the fixed point theory. The discrete time LQG problem of singularly perturbed system has not been studied in the literature yet, despite of the extensive study of the corresponding deterministic counterpart. In this section we will resolve that problem by using the results obtained in Sections 8.1 and 8.2. In addition, we will exploit some ideas developed in Chapter 8 for the LQG problem of another class of small parameter problems - weakly coupled systems.

This section presents the approach to the decomposition and approximation of the linear quadratic Gaussian (LQG) control problem of singularly perturbed discrete systems by treating the decomposition and approximation tasks separately from each other. The decoupling transformation of (Chang 1972) is used for the exact block diagonalization of the global Kalman filter. The approximate feedback control law is then obtained by approximating the coefficients of the optimal local filters with the accuracy of $O(\epsilon^N)$. The resulting feedback control law is shown to be a near optimal solution of the LQG by studying the corresponding closed loop system as a system driven by white noise. It is shown that the order of approximation of the optimal system trajectories is $O(\epsilon^{N+1/2})$ in the slow variables and $O(\epsilon^N)$ in the case of fast variables. All required coefficients of desired accuracy are easily obtained by using the recursive reduced-order fixed point type numerical techniques developed in Chapter 2 and Section 8.1 and 8.2. Obtained numerical algorithms converge to the required optimal coefficients with the rate of convergence of $O(\epsilon)$. In addition, only low-order subsystems are involved in the algebraic computations and no analyticity requirements are imposed on the system coefficients - which is the standard assumption in the power series expansion method. As a consequence of these, under very mild conditions (coefficients are bounded functions of a small perturbation parameter), in addition to the standard stabilizability-detectability subsystem assumptions, we have achieved the reduction in both off-line and on-line computational requirements.

A real world example, a fifth order discrete model of the steam power (Mahmoud, 1982), is included in this section in other to demonstrate the efficiency of the proposed method.

8.3.2 Linear Quadratic Gaussian Control of Discrete Singularly Perturbed Systems at Staedy State

Consider the discrete linear singularly perturbed stochastic system represented in the fast time scale by (see Appendix 8.2):

$$x_1(n+1) = (I + \varepsilon A_{11})x_1(n) + \varepsilon A_{12}x_2(n) + \varepsilon B_1 u(n) + \varepsilon G_1 w(n) \tag{8.32a}$$

$$x_2(n+1) = A_{21}x_1(n) + A_{22}x_2(n) + B_2 u(n) + G_2 w(n) \tag{8.32b}$$

$$y(n) = C_1 x_1(n) + C_2 x_2(n) + v(n) \tag{8.33}$$

with the performance criterion

$$J = E\left\{\sum_{n=0}^{\infty}\left[z^T(n)z(n) + u^T(n)Ru(n)\right]\right\}. \qquad R > 0 \tag{8.34}$$

where $x_i \in R^{n_i}$, $i = 1, 2$, comprise slow and fast state vectors respectively, $u \in R^m$ is the control input, $y \in R^l$ is the observed output, $w \in R^r$ and $v \in R^l$ are independent zero-mean stationary Gaussian mutually uncorrelated white noise processes with intensities $W > 0$ and $V > 0$, respectively, and $z \in R^s$ is the controlled output given by

$$z(n) = D_1 x_1(n) + D_2 x_2(n) \tag{8.35}$$

All matrices are bounded functions of a small positive parameter ε, having appropriate dimensions.

The optimal control law is given by (Kwakernaak and Sivan, 1972)

$$u(n) = -F\hat{x}(n) \tag{8.36}$$

with

$$\hat{x}(n+1) = A\hat{x}(n) + Bu(n) + K\left[y(n) - C\hat{x}(n)\right] \tag{8.37}$$

where

$$A = \begin{bmatrix} I + \varepsilon A_{11} & \varepsilon A_{12} \\ A_{21} & A_{22} \end{bmatrix}, \quad B = \begin{bmatrix} \varepsilon B_1 \\ B_2 \end{bmatrix}, \quad C = \begin{bmatrix} C_1 & C_2 \end{bmatrix}$$

$$K = \begin{bmatrix} \varepsilon K_1 \\ K_2 \end{bmatrix}, \quad F = \begin{bmatrix} F_1 & F_2 \end{bmatrix}$$

The regulator gain F and the filter gain K are obtained from

$$F = (R + B^T PB)^{-1} B^T PA \tag{8.38}$$

$$K = AQC^T(V + CQC^T)^{-1} \tag{8.39}$$

where P and Q are positive semi-definite stabilizing solutions of the discrete time algebraic regulator and filter Riccati equations, respectively given by

$$P = D^T D + A^T PA - A^T PB(R + B^T PB)^{-1} B^T PA \tag{8.40}$$

$$Q = AQA^T - AQC^T(V + CQC^T)^{-1} CQA^T + GWG^T \tag{8.41}$$

where

$$D = \begin{bmatrix} D_1 & D_2 \end{bmatrix}, \quad G = \begin{bmatrix} \varepsilon G_1 \\ G_2 \end{bmatrix}$$

Due to singularly perturbed structure of the problem matrices the required solutions P and Q in the fast time scale version have the form

$$P = \begin{bmatrix} P_{11}/\varepsilon & P_{12} \\ P_{12}^T & P_{22} \end{bmatrix}, \qquad Q = \begin{bmatrix} \varepsilon Q_{11} & \varepsilon Q_{12} \\ \varepsilon Q_{12}^T & Q_{22} \end{bmatrix} \qquad (8.42)$$

In other to obtain required solutions of (8.40)-(8.41) in terms of the reduced-order problems and overcome the complicated partitioned form of the discrete-time algebraic Riccati equation, we have used the method developed in the previous sections (which is based on the bilinear transformation), to transform the discrete algebraic Riccati equations (8.40)-(8.41) into continuous-time algebraic Riccati equations of the form

$$A_R^T P + P A_R - P S_R P + D_R^T D_R = 0, \qquad S_R = B_R R_R^{-1} B_R^T \qquad (8.43)$$

$$A_F Q + Q A_F^T - Q S_F Q + G_F W_F G_F^T = 0, \qquad S_F = C_F^T V_F^{-1} C_F \qquad (8.44)$$

such that the solutions of (8.40)-(8.41) are equal to the solutions of (8.43) and (8.44), i.e.

$$P = P, \qquad\qquad Q = Q \qquad (8.45)$$

where

$$A_R = I - 2(\Delta_R^{-1})^T$$

$$B_R R_R^{-1} B_R^T = 2(I + A)^{-1} B R^{-1} B^T \Delta_R^{-1}$$

$$D_R^T D_R = 2\Delta_R^{-1} D^T D(I + A)^{-1} \qquad (8.46a)$$

$$\Delta_R = (I + A^T) + D^T D(I + A)^{-1} B R^{-1} B^T$$

and

$$A_F = I - 2(\Delta_F^{-1})$$

$$C_F^T V_F^{-1} C_F = 2(I + A^T)^{-1} C^T V^{-1} C \Delta_F^{-1}$$

$$G_F W_F G_F^T = 2\Delta_F^{-1} G W G^T (I + A^T)^{-1} \qquad (8.46b)$$

$$\Delta_F = (I + A) + G W G^T (I + A^T)^{-1} C^T V^{-1} C$$

It is shown in Section 8.1 that the equations (8.43)-(8.44) preserve the structure of singularly perturbed systems. These equations can be solved in terms of the reduced-order problems very efficiently by using the recursive method developed in Chapter 2, which converges with the rate of convergence of $O(\varepsilon)$ under the following assumption:

Assumption 8.3. The matrix A_{22} has no eigenvalues located at -1.

Under this assumption matrices Δ_R and Δ_F are invertible.

Solutions of (8.43) and (8.44) are found in terms of the reduced-order problems by imposing standard stabilizability-detectability assumptions on subsystems (see Assumption 8.2).

Getting approximate solutions for P and Q in terms of the reduced-order problems will produce saving in off-line computations. However, in the case of stochastic systems, where the additional dynamical system - filter - has to be built, one is particularly interested in the reduction of on-line computations. In this section that will be achieved by the use of the decoupling transformation introduced in (Chang, 1972). The Kalman filter (8.37) is viewed as a system driven by the innovation process (Khalil and Gajic, 1984). However, one might study the filter form when it is driven by both measurements and control. The filter form under consideration is obtained from (8.37) as

$$\hat{x}_1(n+1) = (I + \varepsilon A_{11} - \varepsilon B_1 F_1)\hat{x}_1(n) + \varepsilon(A_{12} - B_1 F_2)\hat{x}_2(n) + \varepsilon K_1 v(n) \qquad (8.47a)$$

$$\hat{x}_2(n+1) = (A_{21} - B_2 F_1)\hat{x}_1(n) + (A_{22} - B_2 F_2)\hat{x}_2(n) + K_2 v(n) \qquad (8.47b)$$

with the innovation process

$$v(n) = y(n) - C_1 \hat{x}_1(n) - C_2 \hat{x}_2(n) \qquad (8.48)$$

The nonsingular state transformation of Chang (1972) will block diagonalize (8.47). That transformation is given by

$$\begin{bmatrix} \hat{\eta}_1(n) \\ \hat{\eta}_2(n) \end{bmatrix} = \begin{bmatrix} I_1 - \varepsilon HL & -\varepsilon H \\ L & I_2 \end{bmatrix} \begin{bmatrix} \hat{x}_1(n) \\ \hat{x}_2(n) \end{bmatrix} = T \begin{bmatrix} \hat{x}_1(n) \\ \hat{x}_2(n) \end{bmatrix} \qquad (8.49)$$

with

$$T^{-1} = \begin{bmatrix} I_1 & \varepsilon H \\ -L & I_2 - \varepsilon LH \end{bmatrix}$$

where matrices L and H satisfy equations

$$\varepsilon La_{11} + (I - a_{22})L + a_{21} - \varepsilon La_{12} L = 0 \qquad (8.50)$$

$$H(I - a_{22} - \varepsilon La_{12}) + \varepsilon(a_{11} - a_{12} L)H + a_{12} = 0 \qquad (8.51)$$

with

$$a_{11} = A_{11} - B_1 F_1, \qquad\qquad a_{12} = A_{12} - B_1 F_2$$

$$a_{21} = A_{21} - B_2 F_1, \qquad\qquad a_{22} = A_{22} - B_2 F_2$$

The optimal feedback control, expressed in the new coordinates, has the form

$$u(n) = -f_1 \hat{\eta}_1(n) - f_2 \hat{\eta}_2(n) \qquad (8.52)$$

with

$$\hat{\eta}_1(n+1) = \alpha_1 \hat{\eta}_1(n) + \varepsilon \beta_1 v(n) \qquad (8.53a)$$

$$\hat{\eta}_2(n+1) = \alpha_2 \hat{\eta}_2(n) + \beta_2 v(n) \qquad (8.53b)$$

where

$$f_1 = F_1 - F_2 L, \qquad\qquad f_2 = F_2 + \varepsilon(F_1 - F_{22} L)H$$

$$\alpha_1 = I + \varepsilon(a_{11} - a_{12} L), \qquad\qquad \alpha_2 = a_{22} + \varepsilon La_{12}$$

$$\beta_1 = K_1 - H(K_2 + \varepsilon LK_1), \qquad\qquad \beta_2 = K_2 + \varepsilon LK_1$$

The innovation process v is now given by

$$v(n) = y(n) - d_1 \hat{\eta}_1(n) - d_2 \hat{\eta}_2(n) \qquad (8.54)$$

where

$$d_1 = C_{11} - \varepsilon C_2 L, \qquad\qquad d_2 = C_2 + \varepsilon(C_1 - C_2 L)H$$

Approximate control law is defined by perturbing coefficients F_i, K_i (i = 1, 2), L and H by $O(\varepsilon^k)$, $k = 1, 2, ...,$ in other words by using k-th approximations for these coefficients, where k stands for the required order of accuracy, that is

$$u^{(k)}(n) = -f_1^{(k)}\hat{\eta}_1^{(k)}(n) - f_2^{(k)}\hat{\eta}_2^{(k)}(n) \tag{8.55}$$

with

$$\hat{\eta}_1^{(k)}(n+1) = \alpha_1^{(k)}\hat{\eta}_1^{(k)}(n) + \varepsilon\beta_1^{(k)}v^{(k)}(n) \tag{8.56a}$$

$$\hat{\eta}_2^{(k)}(n+1) = \alpha_2^{(k)}\hat{\eta}_2^{(k)}(n) + \beta_2^{(k)}v^{(k)}(n) \tag{8.56b}$$

where

$$v^{(k)}(n) = y^{(k)}(n) - d_1^{(k)}\hat{\eta}_1^{(k)}(n) - d_2\hat{\eta}_2^{(k)}(n) \tag{8.57}$$

and

$$f_1^{(k)} = f_1 + O(\varepsilon^k), \qquad\qquad d_1^{(k)} = d_1 + O(\varepsilon^k)$$

$$\beta_1^{(k)} = \beta_1 + O(\varepsilon^k), \qquad\qquad \alpha_1^{(k)} = \alpha_1 + O(\varepsilon^k) \qquad i = 1, 2.$$

The approximate values of $J^{(k)}$ are obtained from the following equations

$$J^{(k)} = E\{\sum_{n=0}^{\infty}[x^{(k)^T}(n)D^TDx^{(k)}(n) + u^{(k)^T}(n)Ru^{(k)}(n)]\}$$

$$= tr\{D^TDq_{11}^{(k)} + f^{(k)^T}Rf^{(k)}q_{22}^{(k)}] \tag{8.58}$$

where

$$q_{11}^{(k)} = Var\{(x_1^{(k)} \; x_2^{(k)})^T\} \text{ and } q_{22}^{(k)} = Var\{(\hat{\eta}_1^{(k)} \; \hat{\eta}_2^{(k)})^T\}$$

$$f^{(k)} = \begin{bmatrix} f_1^{(k)} & f_2^{(k)} \end{bmatrix}$$

Quantities $q_{11}^{(k)}$ and $q_{22}^{(k)}$ can be obtained by studying the variance equation of the following system driven by white noise

$$\begin{bmatrix} x^{(k)}(n+1) \\ \hat{\eta}^{(k)}(n+1) \end{bmatrix} = \begin{bmatrix} A & -Bf^{(k)} \\ \beta^{(k)}C & \alpha^{(k)} - \beta^{(k)}d^{(k)} \end{bmatrix} \begin{bmatrix} x^{(k)}(n) \\ \hat{\eta}^{(k)}(n) \end{bmatrix}$$

$$+ \begin{bmatrix} G & 0 \\ 0 & \beta^{(k)} \end{bmatrix} \begin{bmatrix} w(n) \\ v(n) \end{bmatrix}$$

(8.59)

where

$$\alpha^{(k)} = \begin{bmatrix} \alpha_1^{(k)} & 0 \\ 0 & \alpha_2^{(k)} \end{bmatrix}, \qquad \beta^{(k)} = \begin{bmatrix} \varepsilon\beta_1^{(k)} \\ \beta_2^{(k)} \end{bmatrix}, \quad d^{(k)} = \begin{bmatrix} d_1^{(k)} & d_2^{(k)} \end{bmatrix}$$

Equation (8.59) can be represented in a composite form,

$$r^{(k)}(n+1) = \Lambda^{(k)} r^{(k)}(n) + \Pi^{(k)} \omega(n) \tag{8.60}$$

with obvious definitions for $\Lambda^{(k)}$, $\Pi^{(k)}$, $r^{(k)}(n)$ and $\omega(n)$. The variance of $r^{(k)}(n)$ at steady state denoted by $q^{(k)}$, is given by the discrete algebraic Lyapunov equation (Kwakernaak and Sivan, 1972)

$$q^{(k)} = \Lambda^{(k)} q^{(k)} \Lambda^{(k)^T} + \Pi^{(k)} W \Pi^{(k)^T}, \qquad W = \text{diag}(W, V) \tag{8.61}$$

with $q^{(k)}$ partitioned as

$$q^{(k)} = \begin{bmatrix} q_{11}^{(k)} & q_{12}^{(k)} \\ q_{12}^{(k)^T} & q_{22}^{(k)} \end{bmatrix} \tag{8.62}$$

On other hand, the optimal value of J has the very well known form, (Kwakernaak and Sivan, 1972)

$$J^{opt} = \text{tr}\left[D^T DQ + PK(CQC^T + V)K^T \right] \tag{8.63}$$

where P, Q, F and K are obtained from (8.38)-(8.41).

The near-optimality of the proposed approximate control law (8.55) is established in the following theorem

Theorem 8.1. Let x_1 and x_2 be optimal trajectories and J be the optimal value of the performance criterion. Let $x_1^{(k)}$, $x_2^{(k)}$ and $J^{(k)}$ be corresponding quantities under the approximate control law $u^{(\star)}$ given by (8.55). Then under the condition stated in Assumption 8.1 and the stabilizability-detectability subsystem assumptions, the following hold

$$J^{opt} - J^{(k)} = O(\varepsilon^k) \tag{8.64a}$$

$$\text{Var}\left\{x_1 - x_1^{(k)}\right\} = O(\varepsilon^{2k+1}) \tag{8.64b}$$

$$\text{Var}\left\{x_2 - x_2^{(k)}\right\} = O(\varepsilon^{2k}), \qquad k = 0, 1, 2, \dots . \tag{8.64c}$$

The proof of this theorem is rather lenghty and it is omitted. It follows the ideas of Theorems 1 and 2 from (Khalil and Gajic, 1984). In addition, due to the discrete nature of the problem, the proof of our theorem, utilizes the bilinear transformation from (Power, 1967) which transforms the discrete Lyapunov equation (8.61) into the continuous one and compares it with the corresponding equation under the optimal control law. More about it can be found in (Shen, 1989).

8.3.3 Case Study: Discrete Steam Power System

A real world physical example, a fifth order discrete model of the steam power system (Mahmoud, 1982) demonstrates the efficiency of the proposed method. The problem matrices A and B are given by

$$A = \begin{bmatrix} 0.9150 & 0.0510 & 0.0380 & 0.0150 & 0.0380 \\ -0.0300 & 0.8890 & -0.0005 & 0.0460 & 0.1110 \\ -0.0060 & 0.4680 & 0.2470 & 0.0140 & 0.0480 \\ -0.7150 & -0.0220 & -0.0211 & 0.2400 & -0.0240 \\ -0.1480 & -0.0030 & -0.0040 & 0.0900 & 0.0260 \end{bmatrix}$$

$$B^T = \begin{bmatrix} 0.0098 & 0.1220 & 0.0360 & 0.5620 & 0.1150 \end{bmatrix}$$

Remaining matrices are chosen as

$$C = \begin{bmatrix} 1 & 1 & 0 & 0 & 0 \\ 0 & 0 & 1 & 1 & 1 \end{bmatrix}, \qquad D^T D = \text{diag}\{5, 5, 5, 5, 5\}, \qquad R = I$$

It is assumed that $G = B$, and that white noise intensity matrices are given by

$$W = 5.0, \qquad V_1 = 5.0, \qquad V_2 = 5.0.$$

It is shown (Mahmoud, 1982) that this model possesses the singularly perturbed property with $n_1 = 2$, $n_2 = 3$ and $\varepsilon = 0.264$.

The simulation results are presented in the following table

k	$J^{(k)}$	$J^{(k)} - J^{opt}$
0	13.4918	0.229×10^{-1}
1	13.4825	0.136×10^{-1}
2	13.4700	0.110×10^{-2}
3	13.4695	0.600×10^{-3}
4	13.4690	1.000×10^{-4}
5	13.4689	$< 10^{-4}$
optimal	13.4689	

Table 8.3 Approximate values for the criterion

It can be seen from this table that we have quite rapid convergence to the optimal solution, namely, it justifies the result of Theorem 8.1, that $J^{(k)} - J^{opt} = O(\varepsilon^k)$. Note that $(0.246)^6 = 3 \times 10^{-4}$.

8.4 Conclusions

The near-optimum (up to any desired accuracy) steady state regulators are obtained for the deterministic and stochastic linear singularly perturbed discrete systems. The proposed method reduces considerably the size of required off-line and on-line computations, since it introduces the full parallelism in the design procedure.

Appendix 8.1

$$D_1 = A_{11} - S_{11} \underline{P}_1 - S_{12} \underline{P}_2^T - D_{21} D_3^{-1} D_{22} = D_{11} - D_{21} D_3^{-1} D_{22} \,,$$

$$D_3 = A_{22} - S_{22} \underline{P}_3 \,, \quad D = D_3^{-1} D_{22}$$

$$D_{21} = A_{12} - S_{12} \underline{P}_3 \,, \quad D_{22} = A_{21} - S_{12}^T \underline{P}_1 - S_{22} \underline{P}_2^T$$

$$H_1^{(J,\,J+1)} = A_{11}{}^T P_2^{(J)} - P_1^{(J+1)} S_{11} P_2^{(J)} - P_2^{(J)} S_{12}^T P_2^{(J)}$$
$$- \epsilon(E_1^{(J+1)} S_{12} E_3^{(J+1)} + E_2^{(J)} S_{22} E_2^{(J)})$$

$$H_2^{(J)} = E_1^{(J)} S_{11} E_1^{(J)} + E_1^{(J)} S_{12} E_2^{(J)T} + E_2^{(J)} S_{12}{}^T E_1^{(J)} + E_2^{(J)} S_{22} E_2^{(J)T}$$

$$H_3^{(J)} = - P_2^{(J)T} A_{12} - A_{12}{}^T P_2^{(J)} + \epsilon P_2^{(J)T} S_{11} P_2^{(J)} + \epsilon E_3^{(J)} S_{22} E_3^{(J)}$$
$$+ P_2^{(J)T} S_{12} P_3^{(J)} + P_3^{(J)} S_{12} P_2^{(J)}$$

Appendix 8.2

Consider a continuous time-invariant linear singularly perturbed stochastic system represented in the fast time scale by

$$\dot{x}_1(t) = \epsilon A_1 x_1(t) + \epsilon A_2 x_2(t) + \epsilon B_1 u(t) + \epsilon G_1 w(t)$$

$$\dot{x}_2(t) = A_3 x_1(t) + A_4 x_2(t) + B_2 u(t) + G_2 w(t)$$

(a.1)

where $w(t)$ is a zero-mean stationary white Gaussian noise.

To obtain the discrete-time description of this system, we write

$$x(t_{n+1}) = \Phi(t_{n+1} - t_n)x(t_n) + \left[\int_{t_n}^{t_{n+1}} \Phi(t_{n+1} - t)Bdt\right]u(t_n)$$

$$+ \int_{t_n}^{t_{n+1}} \Phi(t_{n+1} - t)Gw(t)dt$$

(a.2)

where $n = 0, 1, 2, \ldots$, and $\Phi(t_{n+1} - t_n)$ is the transition matrix of the system (a.1). Assuming that $t_{n+1} - t_n =$ constant $= \Delta$ (sampling period), the equation (a.2) can be written in the form

$$x_d(n+1) = A_d x_d(n) + B_d u_d(n) + G_d w_d(n)$$

(a.3)

where

$$A_d = e^{A\Delta} , \qquad B_d = \int_0^{\Delta} e^{At}Bdt$$

and

$$A = \begin{bmatrix} \epsilon A_1 & \epsilon A_2 \\ A_3 & A_4 \end{bmatrix}, \qquad B = \begin{bmatrix} \epsilon B_1 \\ B_2 \end{bmatrix} \qquad , G = \begin{bmatrix} \epsilon G_1 \\ G_2 \end{bmatrix}$$

It is easy to see that A_d and B_d have the form

$$A_d = \begin{bmatrix} I + \varepsilon A_{11} & \varepsilon A_{12} \\ A_{21} & A_{22} \end{bmatrix}, \quad B_d = \begin{bmatrix} \varepsilon B_{11} \\ B_{22} \end{bmatrix}$$

More analysis is needed about the stochastic nature of the $G_d w_d(n)$ term. Obviously, the mean value of $G_d w_d(n)$ is equal to zero. On the other hand, the corresponding variance to $G_d w_d(n)$ has the order of

$$\text{Var}\{G_d w_d(n)\} = \begin{bmatrix} O(\varepsilon^2) & O(\varepsilon) \\ O(\varepsilon) & O(1) \end{bmatrix} \tag{a.4}$$

which can be interpreted as of having

$$G_d = \begin{bmatrix} O(\varepsilon) \\ O(1) \end{bmatrix}, \quad \text{Int}\{w_d(n)\} = W_d = O(1) \tag{a.5}$$

and it justifies the model (8.32) used in this section.

Similarly, we can assume the structure of $G_d w_d(n)$ term as

$$G_d = \begin{bmatrix} O(1) \\ O(1) \end{bmatrix}, \quad \text{Int}\{w_d(n)\} = W_d = \begin{bmatrix} O(\varepsilon^2) & O(\varepsilon) \\ O(\varepsilon) & O(1) \end{bmatrix} \tag{a.6}$$

In Section 8.3 we adopt the structure given in (a.5).

REFERENCES

Arkin Y., and S. Ramakrishnan (1983), "Bounds of the Optimum Quadratic Cost of Structure Constrained Regulators", IEEE Trans. Automatic Control, AC-28, 924-927.

Bar-Ness Y., and A. Halbersberg (1980), "Solution of the Singular Discrete Regulator Problem Using Eigenvector Methods", Int. J. Control, Vol.31, 615-625.

Basar T. (1974), "A Counter Example in Linear-Quadratic Games: Existence of Non-Linear Nash Strategies", J. of Optimazition Theory ond Application, 14, 425-430.

Belanger P., and T. McGillivray (1976), "Computational Experience with the Solution of the Matrix Lyapunov Equtation", IEEE Trans. Automatic Control, AC-21, 799-800.

Bertrand P. (1985), "A Homotopy Algorithm for Solving Coupled Riccati Equations", Optimal Control Applications and Methods, 6, 351-357.

Blankenship G. (1981), "Singularly Perturbed Difference Equations in Optimal Control Problems", IEEE Trans. Automatic Control, Vol. AC-26, 911-917.

Butuzov V. and A. Vasileva (1971), "Differential and Difference Equation Systems with a Small Parameter for the Case in which the Unperturbed (Singular) System is in the Spectrum", J. Differential Equations, Vol.6, 499-510.

Calise A., and D. Moerder (1985), "Optimal Output Feedback Design of Systems will Ill-Conditioned Dynamics", Automatica, 21, 271-276.

Chang K. (1972), "Singular Perturbations of a General Boundary Value Problem", SIAM J. Math. Anal. 3, 520-526.

Chemouil P., and A. Wahdam (1980), "Output Feedback Control of System with Slow and Fast Modes", J. Large Scale Systems, 1, 257-264.

Chow J. et al., (1982), "Time Scale Modeling of Dynamic Networks", Springer-Verlag, Lecture Notes in Control and Information Sciences, Vol. 47, 1982.

Chow J., and P. Kokotović (1976), "A Decomposition of Near-Optimum Regulators for Systems with Slow and Fast Modes", IEEE Trans. Automatic Control, AC-21, 701-705.

Cruz J. Jr., and C. Chen (1971), "Series Solution od Two-Person, Nonzero-Sum, Linear Quadratic Differential Games", J. of Optimatization Theory and Applications, 7, 240-257.

Delacour J., M. Darwish and J. Fantin (1978), "Control Strategies of Large-Scale Power Systems", Int. J. Control, 27, 753-767.

Elgard I. O., and E. C. Fosha (1970), "Optimum Megawatt-Frequency Control of Multiarea Electric Energy Systems", IEEE Trans. Power Apparatus and Systems, PAS-89, 556-563.

Fosha E. C., and I. O. Elgard (1970), "The Megawatt-Frequency Control Problem: A New Approach via Optimal Contol Theory", IEEE Trans. Power Apparatus and Systems, PAS-89, 563-578.

Fossard A., and J. Magni (1980), "Frequential Analysis of Singularly Perturbed Systems with State or Output Control", J. Large Scale Systems, 1, 223-228.

Gajić Z. (1986), "Numerical Fixed Point Solution of Linear Quadratic Gaussian Control Problem for Singularly Perturbed Systems", Int. J. Control, 43, 373-387.

Gajić Z., Dj. Petkovski and N. Harkara (1989), "The Recursive Algorithm for the Optimal Static Output Feedback Control of Linear Singularly Perturbed Systems", IEEE Trans. Automatic Control, AC-34, 465-468.

Gajić Z., and N. Rayavarupu (1989), "The Recursive Methods for Singularly Perturbed and Weakly Coupled Linear Steady State Control Problems", (submited for publication).

Gajić Z., and X. Shen (1989a), "Decoupling Transformation for Weakly Coupled Linear Systems", Int. J. Control, Vol.50, 1515-1521.

Gajić Z., and X. Shen (1989b), "Study of the Discrete Singularly Perturbed Linear-Qudratic Control Problem by a Bilinear Transformation", IEEE Trans. Automatic Control, to appear.

Gajić Z., and X. Shen (1989c), "Parallel Reduced-Order Controllers for Stochastic Linear Singularly Perturbed Discrete Systems", submitted for publication).

Geromel J., and P. Peres (1985), "Decentralized Load-Frequency Control", IEE Proceedings, 132, Pt. D., 225-230.

Grodt T., and Z. Gajić (1988), "The Recursive Reduced Order Numerical Solution of the Singularly Perturbed Differential Riccati Equation", IEEE Trans. Automatic Control, AC-33, 751-754.

Haddad A. (1976), "Linear Filtering of Singularly Perturbed Systems", IEEE Trans. Automatic Control, AC-21, 515-519.

Haddad A., and P. Kokotović (1977), "Stochastic Control of Linear Singularly Perturbed Systems", IEEE Trans. Automatic Control, AC-22, 815-821.

Harkara N., Dj. Petkovski and Z. Gajić (1989), "The Recursive Algorithm for Optimal Output Feedback Control Problem of Linear Weakly Coupled Systems", Int. J. Control, Vol.50, 1-11.

Hemker P. (1983), "Numerical Aspects of Singular Pertubation Problems", in "Asymptotic Analysis II", Lecture Notes in Mathematics, 985, 267-287, Springer, New York.

Hoppensteadt F., and W. Miranker, (1977) "Multitime Methods for Systems of Difference Equations", Studies Appl. Math. Vol.56, 273-298.

Ishimatsu T., A. Mohri and M. Takata (1975), "Optimization of Weakly Coupled Systems by a Two-Level Method", Int. J. Control, 22, 877-882.

IEEE Committee Report (1968). "Computer Representation of Exitation System", IEEE Trans. Power Apparatus and Systems, PAS-87, 1460-1466.

Jamshidi M. (1980), "An Overview on the Solution of the Algebraic Matrix Riccati Equation and Related Problems", J. Large Scale Systems, 167-192.

Kato T. (1980), "Pertubation Theory of Linear Operators", Springer-Verlag, New York.

Kautsky J., N. Nichols and P. Van Douren, (1985), "Robust Pole Assignment in Linear State Feedback", Int. J. Control, Vol.41, 1129-1155.

Kenney C., and R. Leipnik (1985), "Numerical Integration of the Differential Matrix Riccati Equation", IEEE Trans. Automatic Control, AC-30, 962-970.

Khalil H. (1980), "Approximation of Nash Strategies", IEEE Trans. Automatic Control, AC-25, 247-250.

Khalil H. (1981), "On the Robustness of Output Feedback Control Methods to Modeling Errors", IEEE Trans. Automatic Control, AC-26, 524-526.

Khalil H. (1987), "Output Feedback Control of Linear Two-Time Scale Systems", IEEE, Trans. Automatic Control, AC-32, 784-792.

Khalil H., and Z. Gajić (1984), "Near Optimum Regulators for Stochastic Linear Singularly Perturbed Systems", IEEE Trans. Automatic Control, AC-29, 531-541.

Khalil H., and P. Kokotović (1978), "Control Strategies for Decision Makers Using Different Models of the Same System", IEEE Trans. Automatic Control, AC-23, 289-298.

Khorasani K. and M. Azimi-Sadjadi (1987), "Feedback Control of Two–Time Scale Block Implemented Discrete-Time Systems", IEEE Trans. Automatic Control, Vol. AC-32, 69-73.

Kokotović P., J. Allemong, J. Winkelman, and J. Chow (1980), "Singular Perturbations and Iterative Separation of the Time Scales", Automatica, 16, 23-33.

Kokotović P., and H. Khalil (1986), "Singular Perturbations in Systems and Control" IEEE Press.

Kokotović P., H. Khalil and J. O'Reilly (1986), "Singular Perturbation Methods in Control: Analysis and Design", Academic Press.

Kokotović P., W Perkins, J. Cruz Jr., and D'Ans (1969), "ε-Coupling for Near-Optimum Design od Large Scale Linear Systems", Proceeding IEE, 116, 889-992.

Kokotović P., and G. Singh (1971), "Optimization of Coupled Nonlinear Systems", Int. J. Control, 14, 51-64.

Kokotović P., and R. Yackel (1972), "Singular Pertubation of Linear Regulators: Basic Theorems", IEEE Trans. Automatic Control, AC-17, 29-37.

Kondo R., and K. Furuta, (1986), "On the Bilinear Transformatior. of Riccati Equations", IEEE Trans. Automatic Control, AC-31, 50-54.

Kučera V. (1972), "A Contribution to Matrix Quadratic Equations", IEEE Trans. Automatic Control, AC-17, 344-347.

Kwakernaak H., and R. Sivan (1972), "Linear Optimal Contro: Systems", Wiley-Interscience, New York.

Lancaster P., and M. Tismenetsky (1985), "The Theory of Matrices", Academic Press, Orlando.

Lapidus L. and N. R. Amundson (1950), "Stagewise Absorption and Extraction Equilibrium", Ind. Engng. Chem. 42, 1071-1076.

Lapidus et.al. (1961), "Optimatization of Process Perfomance". A.I.Ch.E.I. 7, 288-294.

Lee H. (1989), "Recursive Reduced-Order Approach to the Differential Games with Small Parameters", Ph. D. Dissertation in progress, Rutgers University.

Levine W., and M. Athans (1970), "On the Determination of the Optimal Constant Output Feedback Gains for Linear Multivariable Sysems", IEEE Trans. Automatic Control, AC-15, 44-48.

Levine W., T. Johnson and M. Athans (1971). "Optimal Limited State Variable Feedback Cotrollers for Linear Systems", IEEE Trans. Automatic Control, AC-16, 785-793.

Li T-Y., and Z Gajić (1989), "An Iterative Method for Finding Nonnegative Definite Stabilizing Solutions of Coupled Algebraic Riccati Equations", (submitted for publication).

Litkouhi B. (1983), "Sampled-Data Control of Systems with Slow and Fast Models", Ph. D. Dissertation, Michigen State University.

Litkouhi B., and H. Khalil (1984), "Infinite–Time Regulators for Singularly Perturbed Difference Equations", Int. J. Control, Vol.39, 587-598.

Litkouhi B., and H. Khalil (1985), "Multirate and Composite Control of Two-Time-Scale Discrete Systems", IEEE Trans. Automatic Control, Vol. AC-30, 645-651.

Mahmoud M. (1986), "Stabilization of Discrete Systems with Multiple-Time Scales", IEEE Trans. Automatic Contorl, Vol. AC-31, 159-162.

Mahmoud M. (1978), "A Quantitive Comparson Between Two Decentralized Control Approaches", Int. J. Control, 28, 261-275.

Mahmoud M. (1982), "Order Reduction and Control of Discrete Systems", Proc. IEE, Vol.129, Pt.D. 129-135.

Mahmoud M., Y. Chen and M. Singh (1986), "Discrete Two-Time-Scale Systems", Int. J. Systems Science, Vol.17, 1187-1207.

Makila P., and H. Toivonen (1987), "Computational Methods for Parametric LQ Problems - A Survey", IEEE Trans. Automatic Control, AC-32, 658-671.

Mendel J. (1974), "A Concise Derivation of Optimal Limited State Feedback Gains", IEEE Trans. Automatic Control, 19, 447-448.

Miranker W. (1981), "Numerical Methods for Stiff Equations", D. Reidel Publishing Company, Holland.

Moerder D., and A. Calise (1985a), "Convergence of a Numerical Algorithm for Calculating Optimal Output Feedback Gains", IEEE Trans. Automatic Control, AC-30, 900-903.

Moerder D., and A. Calise (1985b), "Two-Time Scale Stabilization of Systems with Output Feedback", J. Guidance, 8, 731-736.

Molen C., and C. van Loan, (1978), "Nineteen Dubious Ways to Compute the Exponential of a Matrix", SIAM Review, 20, 801-836.

Naidu D., and Rao (1985), "Singular Perturbation Analysis of Discrete Control Systems", Lecture Notes in Mathematics, Vol. 1154, Springer Verlag, Berlin.

Oloomi H., and M. Sawan, (1987), "The Observer-Based Controller Design of Discrete-Time Singularly Perturbed Systems", IEEE Trans. Automatic Control, Vol. AC-32, 246-248.

Ortega J., and W. Rheinboldt (1970), "Iterative Solution of Nonlinear Equations on Several Variables", Academic Press, New York.

Ozguner U., and W. Perkins (1977), "A Series Solution to the Nash Strategies for Large Scale Interconnected Systems", Automatica, 13, 313-315.

Papavassilopulos G., J. Medanić, and J. Cruz Jr. (1979), "On the Existence of Nash Strategies and Solutions to Coupled Riccati Equations in Linear-Quadratic Games", J. of Optimization Theory and Applications, 28, 49-75.

Papavassilopoulos G., and P. Olsder (1984), "On the Linear-Quadratic Closed-Loop, No Memory Nash Games", J. of Optimization Theory and Applications, 42, 551-560.

Petkov P., N. Christov, and M. Konstantinov (1986), "A Computational Algorithm for Pole Assignment of Linear Multiinput Systems", IEEE Trans. Automatic Control, Vol. AC-31, 1044-1047.

Petkovski Dj. (1981), "Design of Decentralized Proportional-Plus-Integral Controllers for Multivariable Systems", Computers and Chemical Engineering, 5, 51-56.

Petkovski Dj., and M. Rakić (1978), "On the Calculation of Optimum Feedback Gains for Output Constrained Regulators", IEEE Trans. Automatic Control, AC-23, 760.

Petkovski Dj., and M. Rakić (1979), "A Series Solution of Feedback Gains for Output Constrained Regulators", Int. J. Control, 30, 661-669.

Petrović B., and Z. Gajić (1988), "Recursive Solution of Linear-Quadratic Nash Games for Weakly Interconnected Systems", J. Optimization Theory and Applications, 56, 463-477.

Phillips R. (1980), "Reduced Order Modeling and Control of Two Time Scale Discrete Control Systems", Int. J. Control, Vol. 31, 761-780.

Power H. (1967), "Equivalence of Lyapunov Matrix Equations for Continuous and Discrete Systems", Electronic Letters, Vol.3, 83.

Sezar M., and D. Šiljak (1986), "Nested ε-Decomposition and Clustering of Complex Systems", Automatica, 22, 321-331.

Shen X., and Z. Gajić (1989a), "Near-Optimum Steady State Regulators for Stochastic Linear Weakly Couples Systems", Automatica, to appear.

Shen X., and Z. Gajić (1989b), "Optimal Reduced Order Solution of the Weakly Coupled Discrete Riccati Equation", IEEE Trans. Automatic Control, to appear.

Shen X., and Z. Gajić (1989c), "Near-Optimum Steady State Regulators for Stochastic Linear Weakly Coupled Discrete Systems", (submitted for publication).

Shen X. (1989), "Near-Optimum Reduced-Order Stochastic Control of Linear Discrete and Continuous Systems with Small Parameters", Ph.D Dissertation, Rutgers University.

Starr A., and Y. Ho (1989) "Nonzero-Sum Differential Games", J. Optimization Theory and Applications, 3, 49-79.

Stewart (1973), "Introduction to Matrix Computations", Academic Press.

Su W-C., and Z. Gajić (1989), "Reduced-Order Solution to the Finite Time Optimal Control Problems of Linear Weakly Coupled Systems", (submitted for publication).

Teneketzis D., and N. Sandell (1977), "Linear Regulator Design for Stochastic Systems by Multiple Time-Scale Method", IEEE Trans. Automatic Control, AC-22, 615-621.

Toivonen H. (1985), "A Globally Convergent Algorithm for the Optimal Constant Output Feedback Problem", Int. J. Control, 41, 1589-1599.

Washburn H., and J. Mendel (1980), "Multistage Estimation of Dynamical and Weakly Coupled Systems in Continous-Time Linear Systems", IEEE Trans. Automatic Control, AC-25, 71-76.

West P., S. Bignulac and W. Perkins (1985), "L-A-S: A Computer-Aided Control System Design Language", in "Computer-Aided Systems Engineering", Edited by M. Jamshidi and C. Herget, North-Holland, Amsterdam.

Wilde R., and P. Kokotović (1972), "A Dichotomy in Linear Control Theory", IEEE Trans. Automatic Control, AC-16, 382-283.

Wonham W. (1968), "On a Matrix Riccati Equation of Stochastic Control", SIAM J. Control, 6, 681-197.

Yackel R., and P. Kokotović (1973), "A Boundary Layer Method for the Matrix Riccati Equation", IEEE Trans. Automatic Control, AC-17, 17-24.

Zangwill W., and C. Garcia (1981), "Pathways to Solutions, Fixed Points and Equilibria", Prentice-Hall.

INDEX

Lecture Notes in Control and Information Sciences

Edited by M. Thoma and A. Wyner

Lecture Notes in Control and Information Sciences

Edited by M. Thoma and A. Wyner

Lecture Notes in Control and Information Sciences

Edited by M. Thoma and A. Wyner